TOPICS IN MEASUREMENT

MOORE PROGRAMMED EDUCATION SERIES

William Moore, Consulting Editor
Bucknell University

DICK AND HAGERTY
Topics in Measurement: Reliability and Validity

TOPICS IN MEASUREMENT

Reliability and Validity

WALTER DICK

NANCY HAGERTY

Florida State University

McGRAW-HILL BOOK COMPANY

New York St. Louis San Francisco Düsseldorf
London Mexico Panama Rio de Janeiro
Singapore Sydney Toronto

TOPICS IN MEASUREMENT
Reliability and Validity

Library of Congress Catalog Card Number 78-132341

1 2 3 4 5 6 7 8 9 0 V B V B 7 9 8 7 6 5 4 3 2 1 0

This book was set in Palatino by Black Dot, Inc., and printed on permanent paper and bound by Vail-Ballou Press, Inc. The designer was Paula Tuerk; the drawings were done by Joseph Buchner. The editors were Nat LaMar and Barry Benjamin. Sally Ellyson supervised production.

PREFACE

It would be naive to assume that the reader has no understanding or familiarity with the terms reliability and validity. These terms are quite often used in our everyday speech. However, they have very specific meanings in the field of measurement. It will be the purpose of this short book to familiarize the reader with the various interpretations of these terms: how they may be measured and evaluated, and how various factors may influence their measurement.

If the reader has had a general statistics course, he will find that the topics in this book will build on his understanding of variance and correlation. Review exercises are provided, as well as references to other texts, for those readers who have had no statistics courses.

The objectives of this book may be summarized in two basic statements: (1) After completing this book, the student should be able to continue in the study, on a more detailed and advanced level, of the concepts of reliability and validity; that is, he should be able to utilize successfully the basic vocabulary and conceptualizations which are necessary for the more detailed analysis of the psychometric properties of tests. (2) On completing this book, the student should be capable of evaluating the reported reliabilities and validities of commercial or standardized tests in terms of their adequacy and relevance for his particular needs; that is, he should be able to evaluate the indices of reliability and validity which are reported in test manuals in terms of the manner in which the data were collected, the group of students from whom the data were collected, the testing conditions, and the type (or types) of reliability and validity which is reported. The performance criteria for these objectives are represented by questions which appear throughout the program and by the tests which accompany the program book.

The authors wish to acknowledge the time and effort spent by a variety of students at Florida State University and Bucknell University in the field testing of the book. Data on their performance and their comments were invaluable in the revision process. The authors also greatfully acknowledge the typing assistance of the secretaries in the Computer-Assisted Instruction Center at Florida State University, including Mrs. Ann Welton, Mrs. Louise Crowell, and Mrs. Mary Calhoun.

Walter Dick

Nancy Hagerty

TO THE STUDENT

This text has been developed on the basis of the concept of adjunct autoinstruction. "Autoinstruction" suggests that it is a self-contained text which should provide you with the initial skills related to the topics of reliability and validity. The adjective "adjunct" is used to indicate that the text includes many questions which require you to write in an answer. These questions serve to highlight the important topics from the preceding section and to help you determine if you have mastered the various concepts. Research has also indicated that this approach to instruction should facilitate your retention of the ideas which have been presented.

Each section is one or two pages long. When you reach the questions, you should place a piece of paper over the questions and slide it down until you can read the first question. Stop and write in your answer. Slide the paper down to see the correct answer. If you are correct, go on to the next question. If you are wrong, refer back to the text for an elaboration on the answer.

The appendixes have been especially written to improve your use of the text. For example, if you need a review of the topics of square roots or correlation, brief reviews are provided in Appendixes B and C, respectively. A great number of symbols are used throughout the text; they are summarized for your review in Appendix A. A special analysis-of-variance approach to reliability is presented in Appendix D. And finally, a series of exercises are provided in Appendix E. These exercises are designed to provide additional practice with the various procedures which are taught in the text.

This text has been revised based upon the data gathered from a number of groups of students. The average range of reported times to complete the text has been from eight to twelve hours.

CONTENTS

PART 1

Reliability

1

Introduction to Reliability

One of the dictionary definitions of the term *reliability* is "trustworthiness." The reliability coefficient of a test does, in fact, indicate the trustworthiness of the scores on that test. This trustworthiness is usually expressed in terms of the stability and consistency of the test scores. The terms *stability* and *consistency* are used very precisely with respect to particular measurement procedures which are utilized to compute the reliability of a set of test scores. Notice that a reliability coefficient which is computed for a test refers not to the reliability of the test per se, but to the scores obtained on that test.

Two questions which often arise in a testing situation are: "Is the score which I have just obtained for student x the same score I would obtain if I tested him tomorrow and the next day and the next day?" A reliability coefficient, computed in a certain manner, reflects whether the obtained score is a stable indication of the student's performance on this particular test. The second question is: "Is this test score, which I have just obtained on student x, an accurate indication of his 'real' ability?" The reliability coefficient may be used to estimate the accuracy with which student x's "true score" has been approximated on the test he has just taken.

A reliability coefficient is always represented by a numerical value between zero and one which reflects the stability or consistency of the measurements obtained from a test. To compute the reliability coefficient, (1) a test may be administered twice (with the administrations separated by some interval of time), (2) an alternative form of the test may be administered after a period of time, or (3) a test may be simply administered once. These are the three basic procedures for measuring reliability. These are the kinds of reliability which are most often reported in the manuals for standardized tests.

1. The ______________ of a set of test scores indicates either the stability or consistency of those scores.

---

reliability

---

2. A reliability coefficient reflects which of the following?

____ a. The stability of a set of scores from various tests

____ b. The stability of a set of scores over a period of time

---

b

---

3. A reliability coefficient also reflects the accuracy of a set of scores. Which of the following is the better example of this type of reliability?

____ a. How accurately a set of scores indicates what the students "really" know

____ b. How accurately a set of test items measures a particular subject-matter area

---

a

---

4. Reliability coefficients are expressed as numerical values. If you found one which was less than ______________ or greater than ______________, it would be meaningless.

---

zero, one

---

5. A test must always be administered twice in order to compute a reliability coefficient. (true or false)

---

false

---

6. Two different forms of a test may be administered in order to compute a single reliability coefficient. (true or false)

---

true

---

7. The reliability coefficient for a set of scores on a test indicates ______________.

---

the accuracy or stability of the scores

---

8. The three basic testing procedures for obtaining a reliability coefficient are ______________.

---

a single administration, two administrations of the same test, and administrations of alternative forms of a test

---

Before turning to the actual measurement procedures for obtaining test reliability, it is important to take a brief look at the theoretical basis for the concept of reliability.

Variance and Standard Deviation

Table 1 indicates a series of hypothetical scores obtained on a test by a number of students. The mean and variance of these scores are shown. You will recall that the variance of a set of test scores (or of any set of numbers) is simply the average of the squared deviations of the scores from the mean.

$$\sigma^2 = \frac{\sum_{i=1}^{N}(X_i - \overline{X})^2}{N}$$ Variance of a set of scores

Using the data in Table 1, the variance may be computed by (1) subtracting each score X from the mean of the set of scores, (2) squaring that deviation score, (3) summing these deviations squared for all students, and (4) dividing this sum by the total number of students. The standard deviation of the set of scores is expressed as the square root of the variance. The computational formula which is usually employed is

$$\sigma^2 = \frac{\Sigma X^2}{N} - \overline{X}^2$$ Computational formula for variance

$$\sigma = \sqrt{\sigma^2}$$ Standard deviation

where

X = score of one student
X^2 = a score squared; e.g., if $X = 10$, then $X^2 = 100$
ΣX^2 = sum of all students' squared scores
$\overline{X}$ = mean of scores, that is, $\Sigma X/N$
$\overline{X}^2$ = mean squared; e.g., if $\overline{X} = 5$, then $\overline{X}^2 = 25$
d^2 = square of the difference between a score and the mean for that set of scores

TABLE 1 COMPUTATIONS FOR THE MEAN, VARIANCE, AND STANDARD DEVIATION FOR A SAMPLE SET OF SCORES

X	X^2	$X - \overline{X}$	$(X - \overline{X})^2$
27	729	−3	9
34	1,156	4	16
29	841	−1	1
28	784	−2	4
35	1,225	5	25
31	961	1	1
27	729	−3	9
26	676	−4	16
32	1,024	2	4
35	1,225	5	25
29	841	−1	1
27	729	−3	9
$\Sigma X = 360$	$\Sigma X^2 = 10{,}920$	0	$\Sigma = 120$

$N = 12$

$\overline{X} = 30.0$

$\Sigma X^2 = 10{,}920$

$$\sigma^2 = \frac{\Sigma X^2}{N} - \overline{X}^2 = \frac{10{,}920}{12} - (30)^2 = 910 - 900 = 10 \qquad \sigma = 3.16$$

$$\sigma^2 = \frac{\Sigma d^2}{N} = \frac{120}{12} = 10 \qquad \sigma = 3.16$$

Figure 1 shows a frequency distribution of the scores listed in Table 1. One and two standard deviations to either side of the mean of 30.0 are indicated. When dealing with a normally distributed set of scores, 68.26 percent of the scores will fall between plus and minus one standard deviation of the mean. Figure 1 shows that 8 of the 12 scores (67 percent) fall within the range of 27 to 33. In a normal distribution, plus or minus two standard deviations will include 95 percent of the scores; Fig. 1 indicates that all 12 of the sample test scores fall within this range. Plus or minus three standard deviations from the mean, in a normal distribution, will include 99+ percent of the scores.

The important point of this discussion of variances and standard deviations is this: If the hypothetical students whose scores are shown in

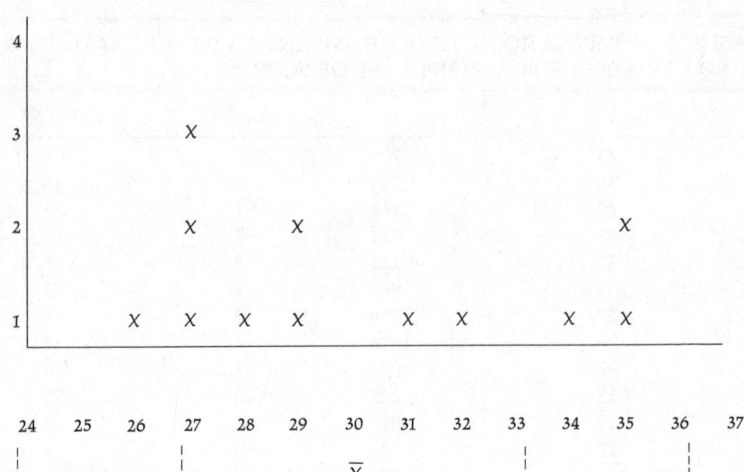

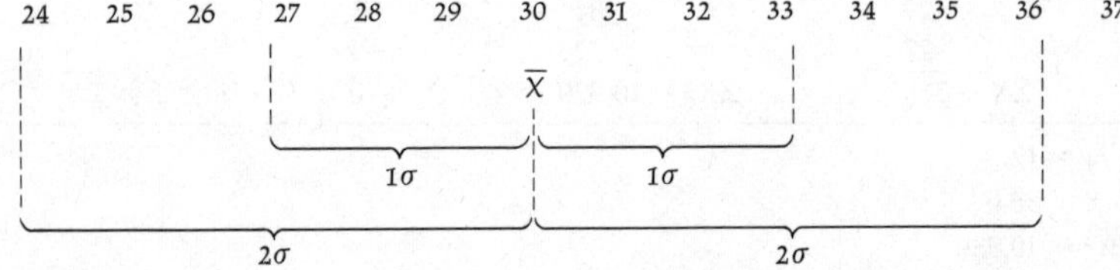

FIGURE 1. Frequency distribution, with mean and standard deviations, for scores from Table 1.

Table 1 had exactly the same amount of ability in performing the task which was being measured, and if the test which was used was a perfect measuring instrument, then all the scores should be identical.

In reality, seldom, if ever, do all the students taking a test obtain the same score, nor is it likely that they have exactly the same ability. Their scores are often distributed like those shown in Fig. 1. The variance of the scores indicates the magnitude of the deviations of a set of obtained scores from the mean. Test theory is primarily concerned with the nature of the variations in scores and the factors which may cause such variations.

1. The ________________ of a set of test scores is the average of the squared deviations of the scores from the ________________.

---

variance, mean

---

2. Standard deviation is defined as the ________________ of the variance.

---

square root

---

3. ________________ percent of the scores in a normal distribution should fall within plus or minus two standard deviations of the mean.

95

---

4. If a test had a mean of 50 and a standard deviation of 10, then a score of 65 on that test would fall between ________________ and ________________ standard deviations ________________ the mean.
(above/below)

1,2, above

---

5. If a student's score is between two and three standard deviations below the mean, at least what percent of the rest of the students had scores that were higher? (Assume that the scores are normally distributed.) Remember the $2\frac{1}{2}\%$ in the upper tail of the curve. ________________

$97\frac{1}{2}\%$

For practice in actually computing the variance and standard deviation of a set of scores, refer to exercise 1 in Appendix E. For a more extensive review of the concepts of variance and standard deviation, consult any standard statistics book in psychology or education.

Test Theory

Test theory includes a number of concepts which refer to the partitioning of the variance of a set of scores. The variance of scores around the mean is conceived of as being made up of two basic components: true-score variance and error variance. The basic concept in test theory states that the score obtained by any individual on a test has two components, namely, his true score (the exact measure of his ability) plus the error associated with his score on this particular test. Thus the assumption is made that no score is a perfect measure of a person's ability. The basic equation which represents this concept is

$$\boxed{X_o = X_t + X_e} \quad \text{Basic test theory equation}$$

where

X_o = score obtained by the student
X_t = his true score
X_e = error associated with his score

The equation indicates that the score obtained by any student on a test is a function of both his true ability and the error associated with the particular measurement.

It is not necessary in test theory to assume that a test is measuring a *single* trait or capability of the individual. The true score may in fact be a summation of a number of different traits or capabilities. The true score may also be the weighted sum of several traits.

1. The basic equation in test theory is $X_o = X_t + X_e$. The equation indicates that an obtained score is really a combination of a ________ score and ________ score.

---

true, error

---

2. The exact measure of a student's ability to perform on a test, as indicated by the equation in question 1, is referred to as his ________.

---

true score

---

3. Given the equation $X_o = X_t + X_e$, indicate the equation for the *error* score. ________

$X_e = X_o - X_t$

---

4. Explain the meaning of the equation which appears as the answer to question 3. ________

Any obtained score is composed of both a true-score and an error component.

---

5. In test theory a person's true score on a test always indicates his ability on a single well-defined task. (true or false)

---

false

Test-theory Assumptions

There are three basic assumptions in test theory. These assumptions are related to the nature of the true scores and the way in which true scores and error scores are combined. The first assumption is that the true score is a measure of a stable and continuing characteristic. It is necessary to assume that the student's performance on the test has not been simply a chance indication of a momentary preference, but rather that it is an indication of a stable ability or trait.

The second assumption is that the error component of a score is a randomly determined variable. That is, each test score reflects not only a student's true ability, but also some degree of error or incorrect assessment of the student's ability. This second assumption indicates that the amount of error associated with any particular score is a matter of chance. The assumption that errors are randomly associated with scores implies that there is no correlation between the obtained scores and the error scores. A student with a high score on a test should be just as likely to have a large error associated with his score, either positive or negative, as the person who received the lowest score on the test. Furthermore, the value of the error component may be either positive or negative, and thus it may either raise or lower the obtained score for a student.

Table 2 shows a set of contrived scores which are the algebraic summation of true scores and error scores. Note that the hypothetical true scores, X_t, have been listed from lowest to highest and that the magnitude of the error scores, X_e, varies randomly. The error scores sum to zero and therefore have a mean of zero, but they do have a variance; i.e., not all the error scores have the same value. The variance in this hypothetical example is obtained in the same manner in which any other variance would be ob-

TABLE 2 DISPERSION OF TRUE MEASURES, ERROR COMPONENTS, AND THEIR SUMS, THE TOTAL MEASURES, WITH MEANS, VARIANCE, AND STANDARD DEVIATIONS (Guilford, 1956, p. 437)

	True Measures X_t	*Error Components* X_e	*Total Measures* X_o
	5	−2	3
	15	+2	17
	20	−4	16
	25	−2	23
	25	+2	27
	25	0	25
	25	+10	35
	30	−4	26
	35	−2	33
	45	0	45
$\Sigma =$	250	0	250
$\overline{X} =$	25.0	0.0	25.0
$\Sigma X^2 =$	1,050	152	1,202
$\sigma^2 =$	105.0	15.2	120.2
$\sigma =$	10.2	3.9	11.0

tained. It can be seen, from Table 2, then, that both the true scores and errors have variances. This leads to the third assumption in the theory of test scores. The assumption states that the obtained scores are simply the sum of the true scores and the error scores. From this assumption the second basic formula associated with test theory can be derived: The best estimate of the variance of the total scores is a simple summation of the variance of the true scores plus the variance of the error scores.

$$\boxed{\sigma_o^2 = \sigma_t^2 + \sigma_e^2}$$ Relation of true-score variance and error variance to total variance

It is important to understand the three basic assumptions of test theory. The following questions review these assumptions.

1. The first assumption is that the true score of a student is a representation of a ________________ characteristic of that student.

---

stable, continuing, or consistent

---

2. The major assumption about the error associated with any test score is that it has a ________________ value for all test scores.
(random/constant)

---

random

---

3. Third, it is assumed that the obtained score variance is composed of both ________________ and ________________.

---

true-score variance, error-score variance

---

4. If an obtained score is the sum of the true score plus error, show the equation which indicates the relationship of the variance of a set of scores to the true-score and error variances. ________________

$\sigma_o^2 = \sigma_t^2 + \sigma_e^2$

---

5. The relationship of true-score and error variance could best be described as:

____ a. Additive

____ b. Multiplicative

---

a

---

6. Use the variances in Table 2 to show that, in this hypothetical example, the equation $\sigma_o^2 = \sigma_t^2 + \sigma_e^2$ is true for this example. ________

$120.2 = 105.0 + 15.2$

---

7. You have probably never seen a set of test scores for a class in which every student received the same score. Explain this phenomenon in terms of test theory.

It is unlikely that every student would have exactly the same true score. There is also a random error associated with each score.

The following section should be read with care. It describes the theoretical derivation of the concept of reliability. The equation $\sigma_o^2 = \sigma_t^2 + \sigma_e^2$, which indicates the relationship between true-score and error variance and their additive relationship to form a total variance of test scores, is a key to the development of the test-theory concept of reliability. If the equation is divided through on each side by the total variance σ_o^2, the following equation is the result:

$$1 = \frac{\sigma_t^2}{\sigma_o^2} + \frac{\sigma_e^2}{\sigma_o^2}$$

The equation indicates that the *ratio* of the true-score variance to the total variance plus the ratio of the error variance to the total variance equals 1. The proportions of the total variance represented by the true scores and by the error scores sum to 1. The ratio of the true-score variance to the total variance, or that proportion of the variance in the obtained scores which may be attributed to the true scores, forms the basic definition of reliability.

By substituting r_{tt} (the symbol for the reliability of a test) for σ_t^2/σ_o^2, it can be seen that the equation may be stated: Reliability equals 1 minus the error variance divided by the obtained variance.

$$r_{tt} = 1 - \frac{\sigma_e^2}{\sigma_o^2} \quad \text{Definition of reliability}$$

The proportion of the total variance which is true-score variance may be obtained by simply subtracting the proportion of the variance which is error from 1. The two basic theoretical equations for the reliability coefficient are

$$r_{tt} = \frac{\sigma_t^2}{\sigma_o^2} \qquad r_{tt} = 1 - \frac{\sigma_e^2}{\sigma_o^2}$$

By the use of these equations and the data in Table 2, it is possible to determine the reliability of the hypothetical scores by substituting the values shown there for the variance associated with the various components of the equations.

$$r_{tt} = \frac{105.0}{120.2} = .874$$

or

$$r_{tt} = 1 - \frac{15.2}{120.2} = 1 - .126 = .874$$

Refer to Table 2 and determine for yourself the formula substitutions that have been made.

1. In this section it has been shown that the *proportion* of variance in a set of scores which can be attributed to ________________ plus the proportion attributable to ________________ sums to ________________.

---

true scores, error, 1

---

2. Reliability is defined as the ratio of ______________ variance in a set of scores to the ______________ variance.

---

true; total, or obtained

---

3. The conventional symbol for the reliability of a test is ______________.

---

r_{tt}

---

4. Use the symbol for reliability in an equation for the reliability of a set of scores which uses only true-score variance and total variance.

$$r_{tt} = \frac{\sigma_t^2}{\sigma_o^2}$$

---

5. Use the reliability symbol in an equation for the reliability of a set of scores which uses error variance and total variance. ______________

$$r_{tt} = 1 - \frac{\sigma_e^2}{\sigma_o^2}$$

---

6. If a set of scores had a total variance of 10 and a true-score variance of 6, the reliability of the scores would be ______________.

$r_{tt} = \frac{6}{10} = .60$

---

7. If a set of scores had a total variance of 20 and an error variance of 12, the reliability of the scores would be ________.

$r_{tt} = 1 - \frac{12}{20} = 1 - .60 = .40$

---

8. How would you determine the *true score* for a student on a test?

Before you give too much thought to this question, it should be pointed out that the discussion in the text up to this point has been concerned with test theory and has involved hypothetical data. There are no practical methods for directly measuring a student's true score on a test. Therefore, to estimate the reliability of a set of scores, it is necessary to use certain computational procedures which have been derived from test theory.

Assumptions of Computational Procedures

Before discussing various computational procedures for estimating reliability coefficients, it is desirable to consider the rationale underlying these methods and to specify the factors which contribute to the true-score variance for each method. The three methods for computing reliability coefficients are based on the principle that *two* test scores may be obtained or derived for a set of students on a particular test. The two sets of scores are correlated to determine the reliability of the test, i.e., to determine the amount of variance in the test scores which may be attributed to true differences among individuals.

One of the methods for estimating the reliability of a test is known as the *test-retest* method. The test-retest reliability of a test is simply expressed as the correlation between the scores from two administrations of the same test to the same students. There is no standard duration of time which should separate the two administrations. However, a minimum of one day and perhaps a maximum of one year are boundaries generally considered acceptable for test-retest reliability estimates.

is administered on two separate occasions, a(n) reliability estimate is obtained.

---

different forms of the same test are administered on two difasions, a(n) ______ reliability estimate is obtained.

---

forms

How is the reliability of a test computed when two alternate forms of est are administered to the same group of students? ______

Scores on the two forms are correlated.

---

4. Knowledge of a specific item of information by a student prior to testing contributes to ______ variance with the test-retest method, but contributes to ______ with the alternate forms.

---

true-score, error variance

---

5. The two forms of the test which are used to obtain the alternate forms reliability estimate should measure the ______ *(same/different)* abilities at the ______ *(same/different)* difficulty levels.

---

same, same

The test-retest interval is usually determined on the basis of the nature of the test. For many tests whose scores will be used immediately, an interval of one day to one week is adequate. However, for test scores which will be used for some time in the future (such as an IQ test), a period of one year between the test and retest is not unreasonable.

The test-retest method for measuring reliability is usually considered to be an index of the *stability* of the test. When the same test is administered twice to a group of students, it is assumed that the same general abilities are required on both occasions. Any temporary conditions, such as ill health or different administrative procedures, are considered to be contributors to the error variance in the obtained scores. They tend to create differences in the test scores which are not consistent across both administrations.

It is generally considered that the greatest source of error in the test-retest reliability estimate is caused by changes which occur within the students themselves during the interval between the test and the retest. For all practical purposes, on many tests, the student is a different person after he has taken the test for the first time. He is now aware of the questions that will be asked during the retest and the answers that he has given to them. He may have an opportunity to determine the correct answers to items that were missed on the first administration. When a long interval elapses between the test and the retest, maturational factors also have their differential effects. The longer the interval, the greater the opportunity for any number of other factors (such as instruction) to occur, and therefore the greater the amount of error variance among the scores.

1. When the same test is administered twice in order to determine the reliability of the scores, the ______ method for determining reliability has been used.

---

test-retest

---

2. With the test-retest method, the two sets of scores obtained from the students are __________ in order to determine the reliability of the scores.

---

correlated

---

3. Match each of the tests with the most appropriate test-retest interval: __________.

1. Stanford-Binet intelligence test
2. Gates second-grade reading test

a. 1 day
b. 1 week
c. 1 month
d. 1 year

Your answers to question 3 could vary, according to the way in which the test is to be used. After children reach the age of six, their IQ scores tend to become quite stable, and thus any of the four answer choices would be acceptable. The exact duration would depend upon the specific use of the test scores. For the reading test, any of the first three answers would be acceptable. Choice d could be omitted, since a third-grade test would probably be administered after one year.

---

4. Indicate for each of the following factors whether they would contribute to the true-score or the error variance when the test-retest method (with an interval of one day) was employed:

1. Noise from someone mowing the lawn (error)
2. Reading skill __________
3. Broken pencil __________
4. Headache __________
5. Knowledge of a specific fact prior to the first test __________
6. Knowledge of the correct answer to an item which was missed during the first administration of the test __________

---

error, true, error, error, true, error

---

5. If you were going to deter[...] administering it on two suc[...] not tell the students th[...] How would your dec[...]

This is a good "thought" questio[...]

Because both general ability and spec[...] ferentially among students from test to re[...] mating reliability has been proposed, namely[...]

The procedure for calculating the alternate fo[...] is to correlate the scores obtained by students o[...] the same test. The administration of the two forms [...] some period of time. This method better estimates th[...] ences in the general abilities of the students because th[...] not appear on both tests. However, the estimate is still [...] temporary factors in the environment, by the test adminis[...] temporary conditions of the students, such as boredom or [...] cannot be assumed that these factors will be present on both ad[...] trations of the test, and therefore they contribute to the error var[...] and reduce the estimated reliability of the tests. Because of these con[...] tions, however, the alternate forms method is often used, since these factors influence the actual use of tests. In summary, the alternate forms procedure for computing a reliability coefficient assumes as true-score variance that difference in scores attributable to general and lasting characteristics of the students and includes as error variance that difference attributable to specific skills with specific items on the two test forms.

6. Changes which take place in an individual between the administration of the two forms of a test contribute to ________ variance.

---

error

---

7. Suppose that *two* different forms of a test were given to a group of students and the time interval between the two administrations was only a few minutes.

a. Would the resulting reliability coefficient have any meaning? ________

b. Would this reliability coefficient be expected to be higher or lower than a coefficient derived from the same tests with the same students but separated by a week? ________

c. List some factors which would contribute to true-score variance when the tests were separated by a few minutes but to error variance when separated by a week. ________

---

a. Yes.

b. Higher. Many of the temporary factors would be constant over the short period, and this would contribute to true-score variance.

c. See text.

The third form of reliability estimate is known as the *split-half method,* or perhaps more accurately, the *internal-consistency estimate of reliability.* The first of these split-half methods was based on the proposition that many of the temporary factors which influence the test-retest and alternate forms methods of estimating reliability could be eliminated if a reliability coefficient could be determined from the scores of a *single* administration of a test. It was therefore hypothesized that two scores could be obtained simply by splitting the test into halves. The scores obtained on the first half of the test could be correlated with scores on the second half of the test. However, it is quite clear that even with this procedure, conditions occur during the administration of the test which can specifically influence either early performance or late performance on the test. This effect will increase the variance attributable to error, and thereby reduce the reliability estimate of the test. On the other hand, the elimination of certain factors through the use of split-half methods can result in overestimates of a test's reliability.

Various methods for dividing tests have been suggested in order to obtain the best possible division of items. One of these methods is to determine the score for each student on the odd-numbered test items and then correlate the score with the score on the even-numbered items. Another method is to correlate the scores obtained during odd- and even-numbered time intervals into which the total testing time might be divided. But even under these more refined conditions, error scores will manifest themselves. For example, performance on individual items may be influenced by such temporary factors as attention to the item, interest, or guessing. The latter is often found to be an important factor when the test is quite difficult. In summary, although the split-half procedure avoids some of the error factors associated with test-retest and alternate forms methods, it still has certain limitations based on the method by which the items are divided. Certainly some methods, such as odd items versus even items, are better than others.

1. The split-half estimate of reliability is considered an ______________ ______________ estimate of reliability.

____

internal-consistency

____

2. Internal-consistency reliability estimates are based upon ______________ *(how many)* administration(s) of a test.

____

one

____

3. To utilize the split-half method for estimating reliability, it is necessary to obtain ______________ score(s) on ______________ test(s)
(*how many*) (*how many*)
for each person.

---

two, one

---

4. Any way of splitting the test, e.g., first half versus the second half, odd items versus even, etc., is equally acceptable. (true or false)

---

false

---

5. The split-half reliability estimate will not indicate the stability of the obtained scores over some period of time such as a week. (true or false)

---

true

---

6. Compare the procedures and sources of error in the test-retest, alternate forms, and internal-consistency methods for obtaining reliability estimates.

Compare your answers with the procedures described on pages 18–24.

2

Computational Procedures

Product-Moment Correlations

It is clear that the test-retest and alternate forms reliability coefficients are represented by the correlation between two sets of test scores. The product-moment correlation coefficient represents the reliability of the test scores under the given conditions. The formula for computing the correlation between two sets of scores is

$$r = \frac{N\Sigma XY - \Sigma X \Sigma Y}{\sqrt{[N\Sigma X^2 - (\Sigma X)^2]\,[N\Sigma Y^2 - (\Sigma Y)^2]}} \qquad \text{Product-moment correlation}$$

For practice in computing a correlation coefficient, refer to exercise 2 in Appendix E. If you have never studied the concept of correlation, you should turn to the general review of the concept, in Appendix C.

The split-half method requires the use of an additional formula. When a test is divided into two parts and the scores are correlated, the result is a correlation between scores on tests which have only one-half as many items as were originally administered. If a 20-item test were scored first on the odd items and then on the even items, the correlation between those scores would be based on scores from two *10-item* tests. What is needed is the reliability of the 20-item test. This *cannot* be determined by simply doubling the reliability of a 10-item test. Instead, it requires the use of the Spearman-Brown formula. The general Spearman-Brown formula for the reliability of a test n times as long as the given test is

$$r_{tt} = \frac{nr_{11}}{1 + (n-1)r_{11}} \qquad \text{Spearman-Brown formula}$$

In this formula n is the ratio of the length of the desired test to the length of the present test (length is defined as number of test items), and r_{11} is the already obtained reliability. If the correlation between the scores on the odd items and the scores on the even items were .50, this correlation based on 10 items would be substituted into the Spearman-Brown formula as follows:

$$r_{tt} = \frac{(2)(.50)}{1 + (2-1)(.50)} = \frac{1.0}{1.5} = .67$$

The value 2 has been substituted for n because it was necessary to determine the reliability of the test twice as long as the two 10-item tests used to obtain the original reliability coefficient. The formula indicates that the split-half reliability of the 20-item test is .67.

It is important to remember that any time the split-half method of reliability estimate is utilized, the Spearman-Brown formula must be applied to the correlation coefficient in order to obtain a reliability estimate which is appropriate for the total test length. To review:

$$n = \frac{\text{number of items in desired test}}{\text{number of items in original test}} \qquad r_{11} = \text{reliability of original test}$$

When the split-half method is used to determine two scores for each student on a single test, each score is based on one-half the total number of test items. Therefore the correlation between the split-half scores on a 50-item test is based on two 25-item tests. This correlation equals r_{11}; the original test length upon which r_{11} is based is 25 items; the desired length of the test is 50 items.

The Spearman-Brown formula can also be utilized with reliabilities obtained by the test-retest and alternate forms methods. For example, if one of these methods yielded a reliability of .90 for a 50-item test, and we were interested in determining what the reliability of that test would be if the test were reduced through random selection to 25 items, the data would be substituted in the formula as follows:

$$r_{nn} = \frac{(\frac{1}{2})(.90)}{1 + (\frac{1}{2} - 1)(.90)} = \frac{.45}{.55} = .82$$

Note that in this case $n = \frac{25}{50}$, because the desired length of the test is 25 items and the original total test length is 50 items. Note also that when the correlated scores are based on the total number of test items, then the original number of items is in fact the total number of items. It can be seen that the reliability of the 25-item test in the preceding example would be approximately .82.

1. The reliability estimate for the test-retest and alternate forms procedures is computed through the use of the ______________.

---

product-moment correlation coefficient

---

2. When the split-half method is used, it is necessary first to compute the correlation between the part scores, and then to employ the ______________.

Spearman-Brown formula

---

3. In the formula $r_{tt} = \frac{nr_{11}}{1 + (n - 1)r_{11}}$, define:

$n =$ ______________

$r_{11} =$ ______________

$r_{tt} =$ ______________

$n =$ length (in terms of number of items) in desired test divided by length (in items) of present test

$r_{11} =$ reliability of obtained scores

$r_{tt} =$ reliability of test of desired length

---

4. In the Spearman-Brown formula, n always equals ______________ when the split-half method has been used to get the reliability of a test one-half as long as that desired.

2

---

5. Reliability estimates derived from methods other than split-half may be substituted in the Spearman-Brown formula. (true or false)

---

true

---

6. If a given test were to be reduced by one-third in length (i.e., it is to be two-thirds as long as it presently is) and its present reliability were .90, the value of n in the Spearman-Brown formula (as shown in question 3) would be ______________, and the estimated reliability of the shorter test would be ______________.

$\frac{2}{3}$, .857

EXPLANATION:

Desired length = 2

Present length = 3

$\therefore n = \frac{2}{3}$

$r_{11} = .90$

$$r_{tt} = \frac{(\frac{2}{3})(.90)}{1 + (\frac{2}{3} - 1)(.90)} = \frac{.60}{1 - .30} = \frac{.60}{.70} = .857$$

---

7. Describe various situations in which you might utilize the Spearman-Brown formula. What two basic pieces of data must be obtained before the formula may be employed?

See the text for situations. You must know the reliability of the present test and the ratio of the length of the new test to the old test.

Kuder-Richardson

Another form of internal-consistency reliability measure which is commonly used is one developed by Kuder and Richardson (1937). The derivations of several formulas for estimating the reliability of a test that had been administered only once was stimulated by the problems, already described, which are inherent in the use of the split-half method of estimating reliability.

There are many ways in which a test may be split in order to compute "half-test" scores, which then enter into the computation of a correlation coefficient. For each split, a different reliability coefficient might be obtained. Kuder and Richardson formulated measures of reliability that used item statistics, as opposed to part or total scores, as the basic unit of measurement. The result is a reliability estimate which is equivalent to the average of all possible split-half coefficients.

The basic Kuder-Richardson formula, generally referred to as *formula 20*, or *K-R 20*, is

$$r_{tt} = \left(\frac{k}{k-1}\right)\left(\frac{\sigma_o^2 - \Sigma p_i q_i}{\sigma_o^2}\right) \qquad \text{Kuder-Richardson formula 20}$$

where

k = number of items in the test

p_i = proportion of students responding correctly to item i

$q_i = 1 - p$ (or proportion of students responding incorrectly)

σ_o^2 = test variance

$\Sigma p_i q_i$ = sum of p times q for all items

The term $k/(k-1)$ is a correction factor which permits r_{tt} to equal 1.0.

The term $p_i q_i$ refers to the difficulty of item i multiplied by 1 minus the difficulty of item i (q). Therefore, if 90 percent of a group of students get a test item correct, then the item difficulty p is said to be .9. The value for p_i is then .9, and $q_i = .1$. The value of $p_i q_i$ for that item would be $.9 \times .1$, or .09. In order to determine $\Sigma p_i q_i$, one must sum all the $p_i q_i$ for all the items in a test.

Refer to the data in Table 3, which show the hypothetical performance of six students on an eight-item test. The table indicates that the total variance of the test scores σ_t^2, is 2.25. The total of the item variances may be obtained by summing the $p_i q_i$ products for each item. The $\Sigma p_i q_i$ is 1.63. Therefore the Kuder-Richardson formula 20 reliability can be computed as

$$r_{tt} = \left(\frac{8}{7}\right)\left(\frac{2.25 - 1.63}{2.25}\right) = (1.14)\left(\frac{.62}{2.25}\right) = .314$$

TABLE 3 ITEM PERFORMANCE OF SIX STUDENTS ON AN EIGHT-ITEM TEST

	Item								
Student	*1*	*2*	*3*	*4*	*5*	*6*	*7*	*8*	*Total*
1	1	1	0	1	1	1	1	1	7
2	1	1	1	1	0	0	1	0	5
3	0	1	0	1	1	0	0	1	4
4	1	1	1	0	1	0	0	0	4
5	0	1	1	1	1	1	0	0	5
6	0	0	1	1	0	0	0	0	2
Number of students correct	3	5	4	5	4	2	2	2	27
Proportion correct, p	0.50	0.83	0.67	0.83	0.67	0.33	0.33	0.33	
Proportion incorrect, q	0.50	0.17	0.33	0.17	0.33	0.67	0.67	0.67	
pq	0.25	0.14	0.22	0.14	0.22	0.22	0.22	0.22	
$\Sigma pq = 1.63$	$\overline{x} = 4.50$		$\sigma^2 = 2.25$		$\sigma = 1.50$				

where

$$k = 8 \text{ items}$$

$$\sigma_o^2 = 2.25$$

$$\Sigma p_i q_i = 1.63$$

The basic assumptions of the Kuder-Richardson formulas are that the test items can be scored 1 for correct and 0 for wrong and that the total scores are the sum of the item scores. Note that it would be inappropriate to use the Kuder-Richardson formulas to estimate the reliability of a questionnaire in which the item scores might range from 1 to 5.

1. The ______________ formula 20 is considered an internal-consistency measure of reliability.

Kuder-Richardson

---

2. The Kuder-Richardson formula 20 reliability estimate is equivalent to the average of all possible ______________ estimates.

---

split-half

---

3. To compute reliability with the test-retest, alternate forms, and split-half methods, one must have the ______________ of the students. The basic unit or score which is used in the Kuder-Richardson formula is the students' ______________ scores.

---

test scores or part scores, item

---

4. The letter k in the Kuder-Richardson formula refers to ______________.

---

number of test items

---

5. The letter p in the formula represents ______________, and q can be expressed as ______________.

---

proportion of students passing an item, $1 - p$

---

6. If the two items in a test had item difficulties (p's) of .80 and .60, then Σpq would equal ______________.
(For practice in computing a Kuder-Richardson reliability estimate, refer to exercise 3 in Appendix E.)

0.40
EXPLANATION:

$$p_1 = 0.80$$
$$q_1 = 0.20$$
$$p_1q_1 = 0.16$$
$$p_2 = 0.60$$
$$q_2 = 0.40$$
$$p_2q_2 = 0.24$$
$$\Sigma pq = 0.16 + 0.24$$
$$= 0.40$$

Sometimes individual item statistics are not available, or it may be uneconomical to obtain them. If this is the case, there is an alternative Kuder-Richardson formula that may be employed. In this instance, however, the reliability estimate is a *lower-bound* estimate of the internal-consistency reliability of a test. A lower-bound estimate implies that it is a conservative, or low, estimate of the test reliability.

The K-R 20 formula takes into account the variance of each item in the term Σpq. If it is reasonable to assume that all the test items have approximately the same difficulty, the term Σpq can be replaced by $k\overline{pq}$, where k is still the number of test items, $\overline{p}$ is the mean difficulty of the test items, or the ratio of the total test mean $\overline{x}$ to the total number of items k, and $\overline{q}$ is equal to $1 - \overline{p}$. Therefore the two Kuder-Richardson formulas can be contrasted as follows:

$$r_{tt} = \left(\frac{k}{k-1}\right)\left(\frac{\sigma_t^2 - \Sigma pq}{\sigma_t^2}\right)$$

K-R 20

$$r_{tt} = \left(\frac{k}{k-1}\right)\left(\frac{\sigma_t^2 - k\overline{pq}}{\sigma_t^2}\right)$$

K-R 21

The formula identified as K-R 21 is referred to as Kuder-Richardson formula 21. If you refer again to Table 3, the K-R 21 reliability coefficient can quickly be computed from the data listed. We can obtain $\overline{p}$ by dividing the mean of the test, 4.50, by the number of test items, 8. The result is .56. Then $\overline{q}$ is $1-\overline{p}$, or .44. Using the Kuder-Richardson formula 21, the reliability estimate would be

$$r_{tt} = \left(\frac{8}{7}\right)\left(\frac{2.25 - (8)(.56)(.44)}{2.25}\right)$$

$$= (1.14)\left(\frac{2.25 - 1.97}{2.25}\right) = (1.14)\left(\frac{.28}{2.25}\right) = .141$$

It can be seen that this estimate is substantially lower than that obtained with the Kuder-Richardson formula 20.

1. The major difference in the computational procedures for the Kuder-Richardson formulas 20 and 21 is that formula 20 requires __________, whereas formula 21 requires __________.

item statistics; test mean, or average item difficulty

2. If one chooses to use K-R formula 21, he must assume that all the test items are approximately __________.

equal in difficulty

3. The reliability coefficient obtained by the use of K-R 21 will be a __________ estimate of reliability; i.e., it will be a __________ *(conservative/liberal)* estimate of reliability.

lower-bound, conservative

4. The term Σpq appears in the K-R 20 formula. The substitute for this term in K-R 21 is ________________.

$k\overline{pq}$

Note that, rather than summing all pq, it is necessary only to multiply $\overline{pq}$ by the number of test items.

---

5. The letters $k\overline{pq}$ in the K-R formula 21 represent:

$k =$ ________________

$\overline{p} =$ ________________

$\overline{q} =$ ________________

$k =$ number of test items

$\overline{p} = \frac{\overline{x}}{k} =$ test mean, or ratio of test mean to number of test items,

$\overline{q} = 1 - \overline{p}$

For practice in computing reliability coefficients with the Kuder-Richardson formula 21, refer to exercise 4 in Appendix E.

There is a basic assumption underlying the application of the Kuder-Richardson formulas which states that the test for which the reliability is to be determined is essentially a unifactor test; i.e., all the items in the test are measuring the same characteristic of the individual. Seldom is one measuring a single factor with a test, and to this extent almost all tests are composed of items which measure more than one characteristic. The greater the diversity of test items, in terms of the skills required to determine the correct answers, the lower the correlations of the performance on the various test items. This decrease in the interitem correlations reduces the obtained *internal-consistency* reliability estimate. As indicated before, an additional assumption is that the test items are scored 1 for correct responses and 0 for incorrect responses, and the total score is the sum of the number of items correct.

It will be indicated later in this text that the split-half method of estimating test reliability is inappropriate with a speeded test. The same is true for the Kuder-Richardson formulas. Because a test is speeded, it will cause the interitem correlations to be artificially inflated, and therefore give a biased estimate of the reliability.

1. In the introduction to the topic of reliability, it was indicated that a reliability coefficient may reflect either the stability or accuracy of a set of test scores. In this respect, the Kuder-Richardson method, like the split-half method, yields a measure of internal ________________, while the test-retest and alternate forms methods result in reliability estimates which indicate the ________________ of a set of test scores.

---

consistency, stability

---

2. Since items which are measuring the same skill are more likely to have ________________ interitem correlations than items which measure
(higher/lower)
different skills, a ________________ factor test is more likely to have the higher reliability (other parameters being equal).

---

higher; uni-, or single-

---

3. Could the reliability of the responses to an attitude questionnaire on which the items were scored on a scale from 1 to 5 be computed using the K-R formula 21? (yes or no)

---

no

---

4. $\text{K-R } 20 = \left(\frac{k}{k-1}\right)\left(\frac{\sigma_o^2 - \Sigma pq}{\sigma_o^2}\right)$

After reexamining the Kuder-Richardson formula 20, explain what components of item information are computed to determine the reliability of a set of scores.

---

item difficulty, $1 - p_i$, and the sum of the pq's

Standard Error of Measurement

The three major methods of estimating the reliability of a test have been presented. There is one additional concept which is important in any discussion of reliability, namely, the standard error of measurement (SE_m, or σ_e). After a reliability estimate has been computed for a test, it is quite easy to then determine SE_m. The formula is

$SE_m = \sigma_o\sqrt{1 - r_{tt}}$ Standard error of measurement

where σ_o is the standard deviation of the test scores.

In Table 3, the SE_m, using the K-R 20 estimate, can be shown to be

$$SE_m = 1.50\sqrt{1 - .314} = (1.50)(.83) = 1.24$$

The interpretation of the standard error of measurement is similar to that of the standard deviation of a set of test scores. It may be stated, with a probability of .67, that a student's obtained score does not deviate from his true score by more than plus or minus one standard error of measurement. The probability is .95 for $x_o \pm 2SE_m$ and .99+ for $x_o \pm 3SE_m$. This is *not* to say that the probability is .67 that a student's true score lies in the interval $x_o \pm 1SE_m$. In this latter case, the probability is either 1 or 0; it either lies within that region or it does not.

This interpretation of the SE_m is predicated upon the assumption that errors in measurement are equally distributed throughout the range of test scores. Research has suggested that the computed SE_m for a test might best be interpreted as an *average* value across the score range.

Some measurement specialists have recommended that the SE_m always be reported for a test, and especially for various subgroups that might have been tested. For example, if a test has approximately the same reliability for two groups, but one group has a σ_o of obtained scores of 14 while another has a σ_o of 7, the corresponding SE_m's would be substantially different. For example, if r_{tt} for both groups were .75, then

$$SE_{m(1)} = 14\sqrt{1 - .75} = (14)(.5) = 7$$

$$SE_{m(2)} = 7\sqrt{1 - .75} = (7)(.5) = 3.5$$

The amount of confidence that could be placed in a score on the test obtained by someone in group 2 would be considerably higher than that obtained by someone in group 1.

1. It is important to remember that the formula for the standard error of measurement is ________________.

$SE_m = \sigma_o\sqrt{1 - r_{tt}}$

---

2. In the SE_m formula, σ_o is the ________________ and r_{tt} is ________________.

---

standard deviation of the scores, reliability of the scores

---

3. If a test had a standard deviation of 10 and a reliability of .84, the standard error of measurement would be ________________.

4.0

---

4. One assumption associated with the standard error of measurement is that error of measurement is _______________ distributed throughout the test score range. Research suggests that SE_m might better be interpreted as an _______________ value across the score range.

---

equally, or uniformly; average

---

5. If the standard error of measurement and the standard deviation for a set of test scores were equal, then the reliability of the scores would have to be _______________.

---

zero: ($SE_m = \sigma_o\sqrt{1-0} = \sigma_o \times 1 \qquad SEm = \sigma_o$)

---

6. The smaller the standard error of measurement, the _______________ the confidence one can have in the obtained score.

---

greater, or more

---

7. If a test were administered to two groups, A and B, and the standard deviations of the scores for the groups were approximately equal but the reliability of the scores for group A was slightly higher than that for group B, which group would have the smaller SE_m, A or B? _______________

A

---

8. Explain your answer to question 7.

You could insert sample data in the formula for SE_m.

Factors that Influence the Reliability of a Test

A number of factors may influence the reliability of any test. These factors can be grouped in the following three categories: (1) factors dealing with the nature of the test items and the test itself; (2) the nature of the students who are being tested; and (3) factors related to the administration of the test. The three basic methods for computing a reliability coefficient that have been presented often lead to differing estimates of the reliability of the test scores. The various conditions which can influence the outcome of the computation of a reliability coefficient further emphasize the notion that no test has a single reliability coefficient.

One of the major factors which will affect the reliability of any test is the length of that test. On an intuitive basis, one can see that as the number of items in any particular test is increased, the chance factors which might enter into a score are greatly reduced or balanced out. For example, performance on a three-item multiple-choice test could be greatly influenced by various chance factors which might influence student responses. However, if the three-item test were lengthened to 30 or 40 items, the error sources would have a greater tendency to cancel each other out, and a better estimate of the true scores of the students would be achieved. In addition, it can also be assumed that, as a test is lengthened, the test maker is providing a more adequate sample of items which will measure the trait in question; i.e., the students will have a greater opportunity to display their ability over a wider range of items which tap the ability being measured.

You may already have noted that increasing the length of the test increases its reliability. This assumption is inherent in the use of the Spearman-Brown formula. Although this formula is used most often in the computation of the split-half reliability coefficient, it may also be used to compute the increased reliability due to tripling or adding ten times as many items to a test.

The Spearman-Brown formula may also be used to determine the number of items which must be added to an already existing test in order to increase its reliability to some desired level. The assumption which must be made in order to use the formula is that the items which the test maker wishes to add will cover essentially the same content or measure the same trait as the already existing items, and that the items themselves will have essentially the same psychometric characteristics (for example, $\overline{p}$, correlation with total score, etc.) as those items already in the test.

The usual form of the formula is

$$r_{nn} = \frac{nr_{11}}{1 + (n-1)r_{11}}$$ Spearman-Brown formula

However, this may be rewritten as

$$n = \frac{r_{nn}(1 - r_{tt})}{r_{tt}(1 - r_{nn})}$$ Alternate form of the Spearman-Brown formula

where n is the ratio of the new test length to the old test length, and r_{tt} is the desired reliability of the new test. It should be carefully noted that the result of this formula, n, is not the number of *items* which must be added to (or subtracted from) a test in order to obtain the desired reliability, but rather it is the *ratio* of the number of items needed in the new test to the number of items already in the original test.

This formula may be used regardless of the number of items which appeared on the original form of the test. An example is a test which had a reliability of .50 in its original form. If it is desirable to make the reliability of this test approximately .75, these values are used in the formula above as follows:

$$n = \frac{(.75)(1 - .50)}{(.50)(1 - .75)} = \frac{(.75)(.50)}{(.50)(.25)} = 3$$

In this example it can be seen that, regardless of the length of the original test, it would require a new test with three times as many items to obtain a reliability of .75. If the original test had 15 items, the new test would require 45 items with the same item parameters and covering the same content as those appearing in the original test in order to produce a reliability coefficient of .75.

1. The reliability of any set of scores can be said to be dependent not only upon the method used, but also upon at least three other factors:

1. ________________
2. ________________
3. ________________

test items, or test; administrative conditions; students tested

2. The text should have made it clear that as a test is lengthened, its reliability should ________________. This relationship is inherent in the ________________ formula.

course

increase, Spearman-Brown

3. If the reliability of a test is to be increased by adding more items, these items should have the same ______________ and ______________.

content coverage, psychometric characteristics

4. The result of using the Spearman-Brown formula in its usual form is an estimate of test ________________. When it is rewritten as $n = \frac{r_{nn}(1 - r_{tt})}{r_{tt}(1 - r_{nn})}$, n represents ________________.

reliability, required ratio of the length of a new test to an old test to obtain a desired reliability

5. If a test had a reliability of .48 and one wanted a reliability of .73, how much longer should the test be? Use the formula listed in question 4. ______________

2.92 times longer

---

6. With reference to question 5, if the original test had 27 items, how many *new* items are needed? Be sure to notice that you already have 27 items, to begin with. ______________

52
EXPLANATION:
$(27)(2.92) - 27 = 78.84 - 27 = 51.84$. Round off to 52.

---

7. Demonstrate what the effect would be on the reliability of a 90-item test (which has a reliability of .85) if 30 items were randomly removed from the test.

You should have used the conventional form of the Spearman-Brown formula to indicate that the reliability would be reduced.

---

8. Why might it be desirable to remove 30 items from a quite reliable test?

Primary reasons are to save either time or money.

Speeded Tests

There is a general distinction in the area of tests and measurements between what are called speeded tests and power tests. Although the distinction is not a clear one, it is generally considered that a *power test* is one in which most of the students have time to finish the test, or at least have time to attempt all the items. A typical example of a power test is an achievement test. On the other hand, *speeded tests* are those in which there is a given time limit within which the students must perform and, in general, most of the students *do not* finish. A clear example of the latter is a test of reading speed. It has been suggested that perhaps a rough demarcation line of 75 percent completion will serve to distinguish a power test from a speeded test. That is, if 75 percent or more of the students complete a test, it may be considered to be a power test.

Generally speaking, it is almost always inappropriate to use any type of split-half or Kuder-Richardson internal-consistency measure of reliability when tests are highly speeded. It is reasonably obvious why this is the case. On speeded tests, students often respond correctly to almost every item they attempt, and what differentiates students is mainly the *number* of items they have time to attempt. Therefore, if the scores on the first half of the test are correlated with scores on the second half of the test, nearly everyone would have all items correct on the first half, and there would be variation among the students on the second half. The result would be a relatively low reliability coefficient. Figure 2 graphically shows this condition. The correlation between the two part scores would be very low because there is no *systematic* relationship between the scores obtained on the first half of the test and those obtained on the second half.

If the odd-even type of split were used, the reliability coefficient would be greatly inflated, because the scores for odd items and for even items would be practically identical for each student, and the reliability would approach 1.0. This would be, obviously, a spuriously high reliability. The test-retest method can be utilized to avoid the obvious limitations of internal-consistency reliability estimates of speeded tests; however, the memory and practice effects will be there, and will therefore contribute to differences in performance between the original test and the retest.

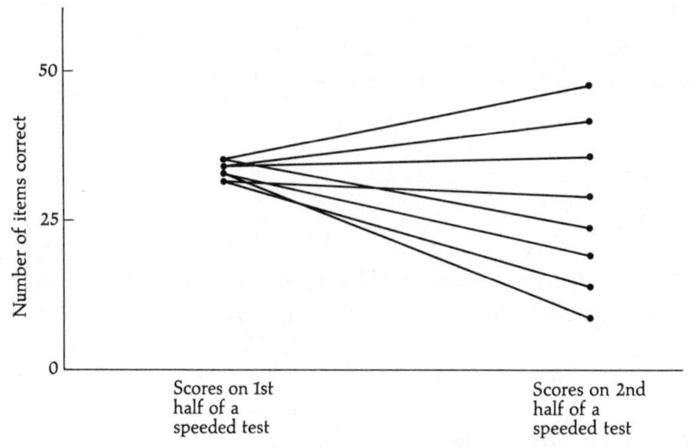

FIGURE 2. Hypothetical distribution of part scores on a speeded test.

The best method for measuring the reliability of a speeded test is to construct two alternative forms which are essentially comparable in terms of their means and standard deviations. These two forms *together* constitute the test for which the reliability is desired. The best tactic is to administer the two alternative forms immediately following each other, with no appreciable time lapse between the two. Thus they are administered as two contiguous but separately timed tests. The correlation between the scores on the two forms may then be inserted in the Spearman-Brown formula in order to estimate the reliability of the entire test.

1. An important distinction is made in testing between those tests which nearly all students have time to complete, namely, ____________ tests, and those tests which relatively few students have time to complete, namely, ____________ tests.

---

power, speeded

---

2. If 15 percent of a group of students completed a test (which is appropriate for them), it would be considered a ______________ test; if 85 percent completed a test, it would be considered a ______________ test.

---

speeded, power

---

3. Kuder-Richardson and split-half reliability estimates are nearly always inappropriate with power tests. (true or false)

---

false
(They are inappropriate with *speeded* tests.)

---

4. The main objection to the test-retest method for estimating the reliability of a speeded test is the influence of ______________ and ______________ effects.

---

memory, practice

---

5. The method recommended for estimating the reliability of a speeded test is the ______________ method.

---

alternate forms

---

6. Note carefully the procedure outlined in the text for using the alternate forms method for estimating the reliability of a speeded test. How does this procedure differ from the usual alternate forms methodology?

There is no time lag between the administration of the two forms.

Restriction in Range of Measurements

Another factor that may significantly affect the obtained reliability coefficient for any test is the nature of the sample of students which is drawn from a specified population for whom the reliability coefficient is to be applicable. An example of such a situation would be the administration of the Stanford-Binet to a group of students who were selected in terms of having very high ability. The result of such a selection procedure would be to reduce significantly the variance of the scores which would be obtained; i.e., there would be relatively little variability among the scores obtained from these students. Furthermore, the reliability coefficient computed from these data would be spuriously low if, in fact, the obtained reliability coefficient were to be generalized to the entire range of abilities measured by the Stanford-Binet.

Alternatively, it is possible to obtain a spuriously high reliability coefficient by including in the test sample students for whom the test is inappropriate. An example of such a situation is one in which a statistics proficiency test is administered both to students who have previously had statistics and to those who have not. Some students will score very high, and a number of students will score almost zero. This will result in a large range of scores, but will not indicate the reliability of the test for the population for which the test is intended. The reliability will be spuriously inflated, if for no other reason than the fact that the item scores for the inexperienced students will nearly all be zeros and thus will have a very high interitem correlation.

A formula has been derived by which the reliability coefficient in one range of the test may be estimated when the reliability of the total test range is known. The major assumption of this formula is that the standard error of measurement is equal for both ranges. The formula is as follows:

$$r_{tt} = 1 - \frac{\sigma_1^2(1 - r_{11})}{\sigma_t^2}$$ Reliability estimate for a new population

where

σ_1^2 = variance of group for which reliability is known
r_{11} = known reliability
σ_t^2 = variance of group for whom reliability is to be determined

This formula may be used as follows: With a test such as the Stanford-Binet, in which the known mean is 100, the standard deviation is 15, and the reliability across all groups is approximately .86, it is possible to estimate what the reliability would be for a selected group which had a standard deviation of scores of only 8. By substituting into the above formula,

$$r_{tt} = 1 - \frac{(15)^2(1 - .86)}{8^2} = 1 - \frac{(225)(.14)}{64} = 1 - .49 = .51$$

The formula indicates that under the condition in which a subgroup has a standard deviation of only slightly more than one-half that of the total population, the reliability of the test for that particular subgroup is reduced from .86 to .51. This is a dramatic example of the effect on reliability of the restriction in the range of abilities of the sample which was selected, in comparison with the range which would be expected in the total population.

1. The examples in the text have shown that when the abilities of a group of students are not representative of the group for whom the test was designed, the resulting reliability estimate made from their scores may be:

____ **a.** Too high
____ **b.** Too low
____ **c.** Too high or too low

---

c

---

2. If the range of abilities in a group that is to be tested is wider than that intended by the test maker, the reliability estimate for this group will be spuriously ______________, whereas if the range of abilities is significantly restricted, the reliability estimate will be spuriously ____________.

---

high, low

---

3. Given $r_{tt} = 1 - \frac{\sigma_1^2(1 - r_{11})}{\sigma_t^2}$, match the terms with their definition:

1. σ_1^2 a. reliability for group with given reliability
2. r_{tt} b. reliability for group for which reliability is unknown
3. σ_t^2 c. variance for group with known reliability
4. r_{11} d. variance for group with unknown reliability

1-c, 2-b, 3-d, 4-a

---

4. A test was administered to a group of students. The standard deviation of their scores was 10, and the reliability estimate was .70. Another group of students was subsequently given the same test, and the standard deviation of their scores was 8. Estimate the reliability of their scores.

.53
(Be sure to convert standard deviations to variances.)

---

5. Explain why a test publisher might be criticized if he reported the reliability of a test for fourth-graders which was based upon scores from students in the third, fourth, and fifth grades.

The reliability would be spuriously high because of the inflated range of abilities in the test group.

Another problem that often arises in relation to the range of ability which is measured by a test is the question of whether an obtained standard error of measurement is applicable across the entire range of scores on the test. Recall that the standard error of measurement is defined as the range on either side of the true score within which two-thirds of the obtained scores are likely to fall. It has been pointed out that the standard error of measurement is a valuable test descriptor, since it indicates the extent of the variation of true scores in the population in terms of raw score units. It is an indication of the degree to which differences which are observed in obtained scores are due to errors of measurement or represent true differences in ability. To utilize the standard-error-of-measurement concept in this manner, one must assume that a test is approximately equal in discriminating power throughout the range of obtained scores. However, there may be reason to suspect that this is not true, especially when some students are responding purely at the chance level.

It may sometimes be advisable to compute the standard error of measurement at a number of points along the range of abilities. This may be done through the use of the computational procedures evolved by Rulon (1939). The rationale for this procedure is based on the ability to derive two scores for each person on the measure in question. A typical method for obtaining the scores would be the odd-even procedure. The formula is

$$SE_m = \sqrt{\frac{\Sigma d^2}{N}}$$ Rulon formula for standard error of measurement

where d is the difference between the scores obtained for an individual on the two halves of the test, and Σd^2 is the sum of the differences squared for all the students in the sample. The difference between the two scores for any person is a measure of the error for that individual; i.e., if the test were perfectly reliable, the two scores would be identical. The error of measurement for all students may be converted into a variance estimate by squaring and summing across all individuals and dividing by n, which produces an estimate of the error variance. This is exactly what the standard error of measurement is—an estimate of the amount of error variance in a set of scores. This procedure may be used at selected ranges within the total population, and the standard errors for these groups compared in order to determine whether, in fact, the standard error is equal within these ranges.

1. In the text it was pointed out that the standard error of measurement can be interpreted for the individual much as the ______________ can be interpreted for a group.

---

standard deviation

---

2. In theory, the error associated with any one test score should be equal to the error associated with any other score on the same test. In reality, there is probably more error associated with very ______________ *(high, low)* scores because of chance factors and with very ______________ *(high, low)* scores because of test difficulty.

---

low, high

---

3. Two test scores (number of odd- and even-numbered items correct) are shown below for eight students. Four students were from the middle of the score distribution, and four from the upper range. By inspection, would you estimate that the SE_m is approximately the same for both groups? Rulon's formula is $SE_m = \sqrt{\frac{\Sigma d^2}{N}}$. ________________

Student	*Middle Range* *Odd*	*Even*
1	8	10
2	7	10
3	9	6
4	14	11

Student	*Upper Range* *Odd*	*Even*
5	16	17
6	20	19
7	18	16
8	19	20

---

no

$SE_m = \sqrt{7.75}$ $SE_m = \sqrt{1.75}$

middle range upper range

---

4. Relate the concept of standard error of measurement to the practice of assigning course letter grades on the basis of arbitrarily looking for "breaks" in a frequency distribution of scores. Or consider the problem associated with the assigning of different grades to two students whose test scores do not differ by one standard error of measurement.

This is another "thought" question!

Item Parameters and Reliability

It has already been pointed out with reference to the Spearman-Brown formula that, given certain conditions, the greater the number of test items, the greater will be the reliability of a test. The formula of course assumes that any items which are added to the test cover the same general content area and have the same general psychometric characteristics. It was also implicit in the discussion of the internal-consistency reliability formulas that the higher the interitem correlations, the higher the reliability of the test. The internal-consistency reliability estimates are related directly to the average interitem correlation. To maximize these interitem correlations, the item difficulties should be approximately equal. Two items which both measure the same trait, but differ greatly in their difficulty, may have a very low interitem correlation. As a general statement, it may be said that the reliability of a test is almost directly related to the degree to which the various test items correlate with each other.

It can also be demonstrated that items with a difficulty of .5 will have maximum variance. Thus an item which is missed by half the students separates the test group into two groups that are maximally dispersed from the item mean (.50). Maximizing this variance is conducive to higher reliability. Although it is true that tests with a large number of items with $p = .50$ will tend to maximize reliability, there are a number of practical considerations. To maximize the discrimination power of a test, i.e., to be able to spread students along the whole score range maximally, it may be desirable to have a large range of item difficulties.

1. The Spearman-Brown formula indicates that, under certain conditions, the greater the number of ________________, the higher the reliability estimate for a test.

---

test items

---

2. The two psychometric properties of test items which will most influence the reliability of the test are item _______________ and interitem _______________.

---

difficulty, correlations

---

3. Test items with a difficulty of _______________ will maximize the variance among the students; however, such items may not provide the _______________ among the scores of students that is desired.

.50, discrimination

---

4. Internal-consistency estimates of reliability are highly dependent upon the _______________; therefore a _______________ *(unifactor, multifactor)* test, other things being equal, would more likely have the higher reliability.

---

interitem correlation, unifactor

---

5. Discuss the pros and cons of choosing test items on the basis of their psychometric properties in order to maximize the reliability of a test.

Items with $p = .50$ will have maximum discrimination, but it may be more desirable to have a whole range of item difficulties.

3

Reliability of Profile Differences and Raters

Two problems which are often of interest in educational psychology are the reliability of differences in profile scores for a student and the reliability of judgments made by a group of raters. Each of these problems will be considered here.

Reliability of Difference Scores

It is sometimes necessary to make judgments about scores on various subtests on a profile, and the question arises as to the reliability of the differences between such scores. In almost every case, the differences between the scores are less reliable than the scores themselves. It may be found that although the subtest scores are quite reliable, they may measure the same thing, and therefore are highly correlated. The high correlation between the two scores tends to reduce the reliability of the difference *between* the scores. Cronbach (1960, p. 287) has provided a formula for estimating the reliability of a difference between two standard scores A and B (standard scores are scores which have been converted to have a mean of 0 and a standard deviation of 1), where A and B are subtest scores on a profile. The formula is

$$r_{\text{diff}} = \frac{r_{AA} + r_{BB} - 2r_{AB}}{2 - 2r_{AB}}$$

Reliability of the difference between two standard scores

where r_{AA} and r_{BB} are the reliabilities of the subtests, and r_{AB} is the correlation between the scores on the two tests. Note in this formula that the r's with similar subscripts represent *reliability*, while the r's with different subscripts (AB) represent the *correlation* between scores on A and B.

The formula can best be understood by inserting various values for subtest reliabilities and intercorrelations. For example, if two subtests each had reliabilities of .85, and the correlation between the two were .70, substitution in the formula would indicate that the reliability of differences between scores on these two subtests would be .50. However, if the subtests were substantially less correlated and less reliable, the reliability of the difference in scores would be essentially the same. For example, for two subtests with reliabilities of .55 and an intercorrelation of .10, the reliability of the differences in scores would also be .50. To maximize the reliability of the difference, it is necessary to maximize the reliability of each of the subparts and to minimize their intercorrelation. Note that, for two subtests with reliabilities of .90 and an intercorrelation of only .10, the resulting reliability of differences between scores on these two subtests would be .90. The explanation for this relationship is this: If two tests are highly correlated, then their reliability is often accounted for in terms of whatever the two tests commonly measure. Thus, for any student, the difference between his scores on these two tests is based primarily on the error variance of the test; the difference score is therefore not very reliable.

If one needs to determine the significance of the difference of scores of two people on the same test, it is possible to utilize the concept of the standard error of measurement. It will be recalled that the standard error of measurement is the standard deviation of the test multiplied by the square root of 1 minus the reliability. If two peoples' scores on the same test differ by at least twice the standard error of measurement, it may be assumed that this large difference between the scores would occur by chance only 1 time in 20, or that this difference (or greater) would occur 95 percent of the time. If the two scores differ by three standard errors, such a difference would be observed 99 percent of the time on repeated testing.

1. It has been shown in certain studies that reading comprehension and reading rate are two highly correlated skills. If a test measured these two skills very reliably, the reliability of the difference between scores on the subject would likely be ______________.

---

low or poor

---

2. If the two subtests in reading were practically uncorrelated and had *low* reliabilities, the reliability of the difference in subtest scores would be ______________.

---

low

---

3. To have maximally reliable differences in subtest scores, the subtests should be highly ______________ and ______________.

---

reliable, uncorrelated

---

4. A student took a test which had four part scores. These test parts had reliabilities as follows: A = .90, B = .90, C = .55, and D = .55. The correlation between tests A and B is .80; the correlation between tests C and D is .10. Would the difference in profile scores for tests A and B be more reliable than the difference in scores on tests C and D?__________

The formula for the reliability of a difference score is

$$r_{\text{diff}} = \frac{r_{AA} + r_{BB} - 2r_{AB}}{2 - 2r_{AB}}$$

Reliability of the difference for tests A and B

$$= \frac{.90 + .90 - (2 \times .80)}{2 - (2 \times .80)} = \frac{1.80 - 1.60}{2 - 1.60} = \frac{.20}{.40} = .50$$

Reliability of the difference for tests C and D

$$= \frac{.55 + .55 - (2 \times .10)}{2 - (2 \times .10)} = \frac{1.10 - .20}{2 - .20} = \frac{.90}{1.80} = .50$$

The reliability of the difference of the scores would be equal.

---

5. If two students obtained scores of 89 and 83 on a classroom test which had a standard deviation of 7 and a reliability of .64, could you justify giving the students different letter grades? ____________________

$SE_m = 4.2$

$(7\sqrt{1 - .64} = 4.2)$

You might be justified in giving different grades because the difference in scores is 6 points, which is at least greater than $1SE_m$. However, you might argue that the difference should be at least $2SE_m$.

Reliability of Raters

There are two basic methods for estimating the reliability of ratings from a number of judges. The first of these is a method developed by Guilford (1954) as a revision of earlier work by Hollingworth (1913). The second is an analysis-of-variance approach, which has been described by Ebel (1951). The first method is applied, basically, to rank ordering of individuals by raters. The second is an analysis-of-variance approach to the intraclass correlation. For the latter, the reader is referred to Ebel's article.

Guilford's formula requires that the k raters rank the N persons from high to low, and then the following formula may be applied:

$$\bar{r} = 1 - \frac{k(4N+2)}{(k-1)(N-1)} + \frac{12\Sigma S^2}{k(k-1)N(N^2-1)} \qquad \text{Guilford's formula}$$

where

$\bar{r}$ = the average intercorrelation among individual judges and, importantly, is therefore the reliability for *one* judge

k = number of judges

N = number of individuals

S = sum of ranks for any individual

Table 4 shows the rank-order ratings for four raters of nine individuals and the accompanying results. Inserting these data into the formula above,

$$\bar{r} = 1 - \frac{(4)(38)}{(3)(8)} + \frac{(12)(4{,}234)}{(4)(3)(9)(80)}$$

$$= 1 - \frac{152}{24} + \frac{50{,}808}{8{,}640} = 1 - 6.333 + 5.881$$

$$= .548$$

This is the reliability for one rater. The Spearman-Brown formula may be used to determine the reliability for the four judges.

$$n \text{ in Spearman-Brown} = \frac{4 \text{ (new length)}}{1 \text{ (old length)}} = 4$$

TABLE 4 DATA TABLE OF RANKINGS OF NINE INDIVIDUALS *(N)* BY FOUR JUDGES *(K)*

Individual	*Judge* 1	2	3	4	*S* *Total Rating*	S^2 *(Rating)*^{2}
1	7	9	6	8	30	900
2	3	1	1	2	7	49
3	2	2	5	1	10	100
4	8	8	9	7	32	1,024
5	5	4	3	4	16	256
6	9	7	8	6	30	900
7	1	5	2	3	11	121
8	6	3	4	7	20	400
9	4	6	7	5	22	484
						$\Sigma S^2 = 4{,}234$

$$r_{tt} = \frac{(4)(.548)}{1 + (3)(.548)} = \frac{2.192}{1 + 1.644} = \frac{2.192}{2.644} = .829$$

Studies have shown that the greatest gains in reliability come from adding raters when the initial reliability of the ratings is quite low. When the number of judges is greater than five or six, the increase in reliability usually is not justified by the addition of more raters.

1. The formula for the reliability of judges' rank ordering of students is

$$\bar{r} = 1 - \frac{K(4N + 2)}{(K - 1)(N - 1)} + \frac{12\Sigma S^2}{K(K - 1)N(N^2 - 1)}$$

Define these components of the formula:

$K =$ ______________

$N =$ ______________

$S^2 =$ ______________

---

number of raters, number of people being rated (sum of a person's ranks)^2

2. The result of applying Guilford's formula to a set of ratings is the reliability of ______________ rater(s).
(one/all)

---

one

---

3. To determine the reliability of the entire set of ratings it would be necessary to employ the ______________ formula, with n in that formula equal to the ______________.

Spearman-Brown, number of raters

---

4. In a study you might have three raters judge students' performance on a certain task. Explain how you would determine the number of raters you might use in a second study which incorporated the same type of ranking.

Using the obtained $\overline{r}$ in the first study, keep incrementing the value of n in the Spearman-Brown formula until the desired reliability estimate is achieved.

For practice in computing the reliability of judges' ratings, refer to exercise 6 of Appendix E.

Relationship of Reliability to the Evaluation of a Test

The heading for this section might well have been "how to either write or interpret the reliability section of a test manual," for it is in the test manual that the reliability coefficient becomes a statistic which is of importance to the test user, a statistic which will be important in the consideration of the usefulness of the test. The following information will pertain mainly to the user's interpretation and evaluation of reliability coefficients, but it will serve equally well as advice to the potential test developer.

The first consideration should be given to the type of reliability reported. An awareness of the three general types of reliability coefficients and their properties is necessary. For example, the test-retest method will result in a reliability coefficient which indicates the relative stability of a set of scores for some group of students over a period of time.

Inspection of the means and the standard deviations on the test and retest forms will indicate what type of change has taken place over that time. The user might also speculate as to what has caused such fluctuations in performance which are observed. The test user might be interested in this test-retest reliability of a test he has already used. The general question the user might ask is: "Are the scores which were taken some time ago on this test still reliable enough for a decision to be made, or should the same test or an alternative form of that test be readministered to better describe the abilities of the students at this time?"

The alternate forms estimate of reliability indicates more the stability of the trait being measured than the stability of the test scores per se, as is indicated by the simple test-retest method. The alternate forms method will generally eliminate memory factors for specific items, but it will not rule out the effect of learning experiences which have occurred during the interpolated time period between the administration of the first and second forms. As indicated before, the alternate forms method is the most appropriate method for estimating the reliability of speeded tests.

Split-half and Kuder-Richardson coefficients should be used to report the internal-consistency reliability of tests. These methods estimate the reliability of the test on a single administration and indicate the extent to which the tests are homogeneous, i.e., measure a single trait. The user should always be wary of the application of internal-consistency methods in the estimate of the reliability of speeded tests. This estimate will invariably be inflated and generally quite inaccurate.

1. If one were to refer to the "precision" of a test, he would probably be referring to a reliability coefficient based on ______________ testing(s).
(one/two)

---

one
(He would probably be referring to *accuracy* rather than stability.)

---

2. The "stability" of a student's test score performance would be indicated best by the ______________ method of estimating reliability.

---

test-retest

---

3. To assess the reliability of the constancy of a trait over a period of time, it would be most suitable to employ the ______________ method for estimating reliability.

---

alternate forms

---

4. On an achievement test in history, probably the most influential factor affecting the retest scores, and therefore the resulting test-retest reliability, is ______________.

---

learning or knowledge of specific items of information or memory

---

5. If a test is speeded, either the ______________ or ______________ method for estimating reliability should be used, preferably the latter.

---

test-retest, alternate forms

---

6. Two methods for estimating reliability from one administration of a test are ______________ and ______________. These two methods are referred to as ______________ estimates of reliability.

---

split-half, Kuder-Richardson, internal-consistency
(If you answered Spearman-Brown, you should be aware that the S-B formula is not used to compute original reliabilities, but to *adjust* reliability estimates for the proper length of the test.)

After inspecting the basic method of estimating reliability which is reported, the user will generally ask the question, "Is the reliability coefficient high enough?" This question has been raised many times, but has seldom been answered. Kelley (1927) has said that, to make decisions about *groups* (such as attitudes of groups, general performance levels of groups, etc.), a reliability coefficient of at least .50 is required. When a decision about a specific individual is to be made on the basis of test results, then a reliability coefficient of .94 or greater should be desired.

In reality, most reliability coefficients fall somewhere between these figures. Some of the best-developed achievement tests generally have reliabilities of approximately .90, whereas classroom-type achievement tests with 50 to 75 items will sometimes have reliabilities as low as .50, and sometimes even lower.

Perhaps the most important point here is that the reliability coefficient should be proportional to the importance of the decision that has to be made on the basis of the scores obtained from the test. For example, the questionnaire to be used to determine the general attitudes of students about classroom practices needs to have only a moderate reliability in order to make certain hypotheses about the students' attitudes. A relatively higher reliability coefficient would be desired of a test which would be used as a college-entrance selection device. Certainly, the highest reliability would be demanded of a test which might be used to determine whether a student should be institutionalized or treated as a severe behavior problem; in fact, several tests should be used for such critical decisions.

1. Although there are no absolute scales for evaluating reliability coefficients, it may be said that the magnitude of the reliability coefficient should be proportional to the ______________ which is to be made.

---

importance of the decision

---

2. Professionally developed standardized tests often have reported reliabilities above ______________, while classroom tests often have reliabilities of ______________ or below.

.90, .50

---

3. Rank-order, from highest to lowest, the following tests in terms of the magnitude of the reliability coefficients which you, the test user, would desire:

____ **a.** Achievement test for course placement
____ **b.** Personality inventory for use in an experiment
____ **c.** Group-attitude questionnaire about food costs
____ **d.** Intelligence test to determine institutionalization

---

d = 1, **a** = 2, **b** = 3, **c** = 4
(Listed above is one possible ranking. Depending upon the user's interpretation of an item such as **b**, the rating scale could change. Certainly, in any case, item **d** should have the highest reliability.)

It should be remembered that there is no single reliability coefficient for a test. The test-retest, alternate forms, and internal-consistency coefficients for a single test may all differ from one another. Additional considerations are the group to whom the test was administered in order to determine the reliability coefficient, and the standard deviation and the standard error of measurement which are associated with their scores.

The persons for whom any test is to be applicable may be divided into many, many groups on the basis of such things as sex, age, occupation, geographic area, etc. Test makers cannot be expected to produce norms for their tests and report reliabilities for all these groups. The test user should be aware of the group (or groups) to whom the test was administered in order to obtain the reliability coefficients which are reported. The more nearly one of these groups resembles that of the user, the more confidence he may have that the same reliability can be achieved with his group.

With respect to the magnitude of the reliability coefficient, the user should be aware that a coefficient of .70 with a rather narrow range of talent in the testing group is fully as good as a coefficient of .90 based on a group with a much greater range of talent. Under certain, not too unusual, circumstances, the standard error of measurement for these two groups could be approximately the same.

Other user precautions should be noted. The first is that the reliability coefficient should never be misconstrued or misrepresented as the validity of the test. A test can be highly reliable but completely invalid for the user's purpose or for the purpose which the test maker claims. The user should also be aware that the reported intercorrelation between raters, or scorers, indicates the extent to which the raters have, in the past, agreed on the performance of students. However, it is not an indication of the reliability of the test itself, but rather of the raters. It is also possible that the scorers' ratings could agree perfectly but the test-retest reliability of the ratings could be almost zero. This could happen if the behavior which was being rated was highly unstable.

1. The test user must be aware that a test may have ________ reliability coefficient(s). In addition to test reliability, the standard deviation and standard ________ should also be reported.

---

more than one or many, error of measurement

---

2. Since test makers cannot produce norms for every conceivable subgroup, the user should check to see if ________.

---

there is a group similar to his

---

3. The test maker who claims that, because his test has a high test-retest reliability it is therefore valid for certain purposes, may be considered to be ________.

---

wrong, incorrect, uninformed, devious, etc.

---

4. Two precautions to the user were indicated in the text in terms of tests for which judges' ratings are reported. The first precaution was related to the fact that the reported judges' reliabilities indicate how judges have rated students ________________, and therefore it is not a guarantee of how reliable the user's judges will be. It was also noted that the judges' reliability could be quite high, and still the test may have no ____________ reliability.

---

in the past, test-retest

Wesman (1952) has developed what he calls a mental checklist, which should be compared with the data which are reported in the reliability section of a test manual. The questions are intended to encourage the user to consider carefully the reliability coefficients which are reported. These questions serve as a useful summary for this discussion of reliability.

1. What does the coefficient measure?
   a. Precision of the test—coefficient based on a single setting?
   b. Stability of examinees' test performance—coefficient based on test-and-retest with a few days intervening?
2. Is it more than a reliability coefficient? Does it also measure constancy of the trait? Is the coefficient based on test-and-retest with enough intervening time for learning or similar changes to have occurred? (Are alternate forms of the test used in the retesting?)[1]
3. Do scores on the test depend largely on how rapidly the examinee can answer the questions? If so, is the reliability coefficient based on a test-retest study? (Using an alternate form.)
4. Are there part scores intended for consideration separately? If so, is each part score reliable enough to warrant my confidence?
5. Is the group on which this coefficient is based appropriate for my purpose? Does it consist of people similar to those with whom I will be using the test?
6. Since the reliability coefficient, like any other statistic, requires a reasonable

[1]The comments in parentheses have been added by the authors.

number of cases to be itself dependable, how large is the group on which the coefficient is based?

If and *only* if the coefficients can be accepted as meeting the above standards may one ask:

7. In view of the importance of the judgments I shall make, is the correlation coefficient (reported reliability coefficient) large enough to warrant my use of the test?

5. Write down a checklist of things to check out when determining the reliability of a test.

The foregoing list of items 1 to 7 is a good model.

PART 2

Validity

4

Introduction to Validity

Suppose a perfectly reliable test were found. Such a test would consistently give the same score for a given individual from one situation to another, from one day to the next. The score would be the student's true score. If the test were given to a group of individuals, the score each one got would be his true score. If he took the test again, at any time under any circumstances, he would get the same score.

A test which was perfectly reliable would seem to be quite valuable, but the test user should also raise the questions: "How valid is it? Does the test measure what I want it to measure?" A perfectly reliable test may not measure anything of value, and it may not correlate with any other test score.

Validity information gives some indication of how well a test measures a given area, under certain circumstances and with a given group. It is for this reason that any one test may have many types of validity, and a unique validity for each circumstance and group tested. All students in grade 1 through graduate school may be given Test XQR. This test may predict success in later life. For each grade, there is a different validity coefficient possible. Also, there is a different one for each group of females. There is another for students entering the engineering curriculum, one for nurses, one for computer programmers.

The author of a test usually lacks the subjects and resources to obtain validity coefficients for all possible groups. It is for this reason that a representative group is tested, and coefficients stated for that group. Users of the test will have to judge for themselves how well their group is represented by the group for whom the validity data were collected.

If the test is to be used over a period of years on a specific college campus, for example, it is advisable for the administrators to obtain validity data on the college groups in question. This will be more accurate for the particular circumstances than reliance on the author's stated validity of the test.

Validity is stated always with reference to the criteria used. A test can have a validity coefficient for each criterion with which it is correlated. This parallels the statement previously made, that each group can have numerous validity coefficients, depending on the test and criterion used. Test XQR can be validated, using as a criterion success in later life. It may also have a separate validity correlation for success as a teacher, for graduation from college, for ability to get along with people, and so forth.

With this in mind, one must be aware that the topic of *validity* may be considered quite theoretical. It is in part a mathematical approach to "proof." As in reliability, the final product is a hypothetical entity, derived from various measured and impossible-to-measure factors.

Types of Validity

There is no *one* type of validity, just as there is not just one type of reliability. Rather, there are several types. Consider the following sample of types of validity:

Content
Concurrent
Construct
Predictive
Face
Curricular
Differential

The first four types (content, concurrent, construct, and predictive) are those with which users of psychological instruments are primarily concerned. The other three are terms coined by various authors to express their needs for other types of validities. And this list is not exhaustive. Cash validity is a term used to refer to the popularity of a test—how much "cash" it brings in. Be aware that other *types* exist, but are not of primary concern in this text.

1. Since reliability and validity are, essentially, theoretical concepts, they constitute a major portion of what is known as the ______________ of measurement.

---

theory

---

2. Recall the four types of validity with which users of psychological instruments are primarily concerned:

1. ______________
2. ______________
3. ______________
4. ______________

---

content
concurrent
construct
predictive

---

3. If a test is said to be valid, we assume that it ______________.

---

measures what we want it to measure

---

4. How many types of validity are there? ______________

Many. (The main concern in this text is with content, concurrent, construct, and predictive.)

Uses of Validity

The type of validity (or validities) which should be estimated for a test is determined by the aims of the testing situation. One tests whether the instrument being validated correlates with a criterion: Do medical students who get high scores on a selected aptitude test also tend to receive the highest final grades in their class? Do those who score lowest also tend to receive the lowest grades in the class? If the correlation between the aptitude scores and the final grades is high, the test does actually predict success in medical school, for this group. The correlation coefficient, in this case, would represent the predictive validity of the test.

Given that a high correlation does exist between test scores and course performance, one can be fairly certain that a group of medical students who pass the test will succeed in medical school. The validity of the test under these circumstances has been established.

This is an example of predictive validity, which will be explained in greater detail later. The use of tests with established *predictive* validity is common in educational, vocational, and psychological areas. An administrator may want to know if the students entering school today will be able to do the required work and receive the necessary grades to be able to graduate in four years. By giving them a previously validated test, he can predict the success or failure of a majority of these students.

For a test to be validated, it must first be administered to a *representative* sample of subjects (e.g., entering freshmen). Then, after some prescribed time has passed, criterion scores (e.g., cumulative grade point averages) must be determined. The correlation of these test scores and grades provides the predictive validity coefficient. If the coefficient is high, it can be utilized to predict the achievement of the students in the next entering class. If it is low, it will be of little value.

This is a brief explanation of one type of validity. Before continuing with the other types, it is important to understand clearly the significance of the *validating group* and the *criterion.*

1. A test which has a high correlation with success in nursing for one group may also be able to ______________ success for other groups entering nursing.

---

predict

---

2. To validate a test, it must be administered to a ______________ group of subjects.

---

representative

---

3. Test results are sometimes correlated with some type of outcome in order to determine the validity of the test. This "outcome" is usually called a ______________.

---

criterion

---

4. Validity coefficients aid in determining how well a test is ______________ the criterion.

---

predicting

---

5. A test which has a high validity coefficient for predicting the success or failure of entering engineering students could also be used to predict the success or failure of any other group of entering students. (true or false)

Perhaps true, but a validation study should be conducted first.

---

6. A test which has a low correlation with the desired criterion should be:

____ **a.** Replaced
____ **b.** Revised
____ **c.** Factor-analyzed
____ **d.** Any of the above

---

d

---

7. How does one decide which type of validity to estimate for a test?

To decide which type of validity estimate to obtain for a test, one must first decide how the test is to be used.

Validation Group

If a test is needed to predict the success of entering college freshmen in terms of successful completion of an engineering curriculum, the representative sample of students in this case must be entering freshmen in the engineering curriculum. Were one to use graduating seniors or secretaries or school teachers as the validating group, the prediction would not be representative of the group to be tested. If the test is intended for underprivileged children, then one should state how "underprivileged" has been or should be determined. It is critical that the potential user of a test be able to evaluate the similarity of his test population to that with which the test was originally validated. If the comparison is too discrepant, then the use of the test for the new population may be quite speculative, at best. The validation group is therefore a group that is representative of the group with which the test is to be used. This group is given the test in order to determine the validity in terms of some criterion already decided upon.

Criterion

One of the problems encountered when attempting to obtain a validity coefficient is the choice of a *criterion.* The example previously used of medical students and final grades is a good case in point.

In this instance, the criterion is the final grade point average. Any student should be aware of the fallacy of using this as the ultimate criterion. Not only do methods of scoring tests differ, but each teacher also gives letter grades using his own standards, and each teacher is capable of using a different standard for each class he teaches. Therefore the composite score is not nearly as valid as the decimal digits might imply.

It is also found that previous performance can contaminate the criterion. For example, if Johnny is an honor student who has received straight A's in school, he could encounter several problems. In one instance, a teacher might feel that Johnny should continue getting A's and give him one whether or not he earned it. In another instance, a teacher might feel that a B on Johnny's record would "do him good." Correlations are biased by the degree to which past performance influences present ratings. Therefore it is essential when validating a test to keep previous records from those who will be making the criterion ratings.

However arbitrary the criterion of grades is, it is one of the most often used when tests are validated. In the educational setting one must recognize the shortcomings of grades if he intends to use them. It should also be noted that judges' ratings and supervisors' ratings suffer from the same types of bias and unreliability. Ideally, the criterion performance should be defined in terms of clearly stated behaviors which can be observed. The agreement of judges as to the presence of such behavior increases the reliability of the criterion scores, and hence the correlation with the "predicting" test.

1. The representative sample used in validating a test is called the ______________ group.

---

validation

---

2. The validation group must be ______________ of the population for whom the test is intended.

---

representative

---

3. The criterion of grade point averages is considered by the authors to be:

___ **a.** The best available
___ **b.** The only criterion to use
___ **c.** The least reliable of all criteria
___ **d.** A fallible, but useful, measure

---

d

---

4. Previous experience with a student can be quite valuable to a teacher in an instructional setting. However, it can also result in a __________ criterion score.

---

contaminated or unreliable or biased

---

5. A test may have more than one validity coefficient. (true or false)

---

true

---

6. Using test X as an example, give a number of ways one could obtain more than one validity coefficient for X.

Correlate with a variety of criteria, e.g., policemen, nurses, and school teachers.

---

7. How does "type" of validity differ from the validity coefficients determined in item 6?

The type refers to the use of the test, e.g., prediction, content, and so on. The validity coefficient is the statistical measure of validity.

5

Types of Validity and Their Uses

Content

The American Psychological Association gives the following definition of content validity:

> The test user wishes to determine how an individual performs at present in a universe of situations that the test situation is claimed to represent.
> (American Psychological Association, Standards for Educational and Psychological Tests and Manuals, 1966, p. 12)

The contents of the test (the test items themselves) are samples of situations. How well those items sample the particular situation(s) under consideration is termed *content validity*. Test constructors have an obligation to justify inclusion of the items on the basis of how well they sample the situation(s) involved, as well as to define the situations themselves. Achievement and ability tests are samples of situations; therefore users of these tests are interested in the content validity.

In a typical achievement test, it is desirable that students be tested on material to which they have had some reasonable exposure. Therefore the test user often scans a potential test in order to identify the content areas and the abilities covered and to estimate the difficulty of the test

TABLE 5 CONTENT MATRIX

	Topic							
	A	B	C	D	E	F	G	*Total*
Recall	2	3	1	1	2	2	3	14
Synthesis	1	2	2	1	2	1	2	11
Evaluation	3	2	1	2	3	2	1	14
Total	6	7	4	4	7	5	6	39

items. This process can be greatly facilitated by the test producer who clearly defines the topics which are tested. The producer should also indicate the types of items employed, e.g., recall, synthesis, or evaluation items. A table such as Table 5 might be used to establish the content validity of the test. (Cell entries represent the number of items in the test which fall in each category.)

Examination of such a table will reveal the topics which are included (and those which have been excluded) from the test and the types of skills which are emphasized within each topic. On the basis of this information, the test user can make an informed judgment of the appropriateness (or content validity) of the test for the user's particular test population.

1. If test items are to be considered valid, they should require that the student perform tasks which are relevant to the total behavior that is being measured. Therefore test items should be examples of various ______________.

---

situations or behaviors

---

2. The extent to which the items in a test do, in fact, sample an area is termed the ______________ ______________ of the test.

---

content validity

---

3. Content validity is quite crucial when one is considering the adoption of a(n) ______________ test.

---

achievement or ability

---

4. Test constructors can justify the inclusion of items in a test by indicating in a matrix: (a) the topics covered in the test and (b) ______________.

---

the type of skill required by the test items

---

5. Which of the following situations requires a test with "content validity"?

____ **a.** Estimate of success before the class is begun
____ **b.** Estimate of success in another field
____ **c.** Estimate of coverage of a particular ability

---

c

Predictive and Concurrent Validity

The test user wishes to forecast an individual's future or to estimate an individual's present standing on some variable of particular significance that is different from the test.

(American Psychological Association, Standards for Educational and Psychological Tests and Manuals, 1966, p. 12)

When one wishes to predict future performance, one is concerned with predictive validity. Tests which significantly correlate with subsequent performance are said to have predictive validity. Validation of this type generally requires long-range, time-oriented analysis.

One might give a college aptitude test to a group of high school seniors. Validation would take place several years later, when these scores are correlated with some criterion, such as final grade point average in college. In this situation, it should be noted that the criterion, grade point average, may be of little value if the students go to a variety of colleges, and a simple "graduate–no graduate" criterion might be more useful. An alternative solution would be to develop a test validity coefficient for each college based upon the scores of the students who attend. This procedure is sometimes utilized with state-wide general-ability examinations.

1. Future performance can be predicted only from the results of a test which has high ______________ validity.

---

predictive

---

2. The criterion used to validate a predictive test is:

____ **a.** Immediately available
____ **b.** Already established
____ **c.** Available in the future

---

c

---

3. Validation of a predictive test involves a time period between testing and the establishment of criterion scores. Is there a fixed time which must elapse between these two events? ________________

no

---

4. If a test does not have a reasonable predictive validity coefficient, it is unwise for the test user to predict ________________ on the basis of its scores.

future performance

---

5. Which of the following situations requires a test with predictive validity?

____ **a.** A classroom spelling test
____ **b.** College entrance exams
____ **c.** College placement tests
____ **d.** Mathematics achievement tests

b

Concurrent validity might be termed "immediate predictive validity," though this is somewhat misleading, because concurrent validity does not predict. To determine concurrent validity, test results are correlated with currently available criterion evidence rather than future information. It is sometimes found that a test with high concurrent validity may not have high predictive validity. Tests which stress concurrent validity are often used to classify patients, differentiate vocational choices, etc.

For example, results of a test which purports to identify mentally disturbed children might be correlated with a group consensus of psychiatrists who examine the same children. The resulting correlation between the test results and the doctors' judgments would be referred to as the concurrent validity of the test.

The test producer who develops a new form of some test often appeals to the user by indicating that the results of his new "two-minute wonder" IQ test correlate .99 with some standardized (and time-consuming) IQ test. The producer is, in fact, using the concurrent validity of his test as its major selling point.

1. "Immediate predictive validity" is correctly called ________________.

---

concurrent validity

---

2. The concurrent validity of a test is determined by correlating scores with other information which is ________________.

---

immediately or currently available

24

3. An example of an immediately available criterion could be a previously validated test on personality characteristics. The *test* for which the validity is sought could be ________________.

---

a new (shorter) personality inventory

---

4. Indicate several situations in which you would be much more interested in the concurrent validity than the predictive validity of a test.

Medical, mental, vocational, or adjustment classification

---

5. Which of the following situations requires a test with concurrent validity?

____ **a.** A mental patient is given an achievement test.
____ **b.** A doctor has a pregnant woman x-rayed.
____ **c.** A doctor gives a mental test to a hospitalized student.
____ **d.** A mental patient is given a personality test prior to dismissal.

---

d

Construct Validity

> The test user wishes to infer the degree to which the individual possesses some hypothetical trait or quality (construct) presumed to be reflected in the test performance.
> (American Psychological Association, Standards for Educational and Psychological Tests and Manuals, 1966, p. 12)

Construct validity deals with the validation of theory. A construct can be considered to be a hypothesis or explanation of some type of behavior. A hypothetical construct such as anxiety or ego might be examined. Neither of these qualities can be seen with the human eye, yet they are often used to explain behavior. Psychologists often develop theories of how people react to situations based on these characteristics. They gather together a group of items they feel should measure that construct. Predictions are made on the basis of the theory as to how individuals will respond to these items.

Data may be gathered by giving a test of "anxiety" to laboratory subjects exposed to anxiety-provoking situations. If the subjects under stress respond to the items as predicted, then the test is purported to measure anxiety. If the subjects do not conform to expectations, the items are revised, or the theory is reexamined.

Although predictive validity can be established for a test by conducting a single study, it often requires numerous studies to identify the constructs which a test is measuring. Procedures used in these studies include the investigation of the particular interrelationship of the test items (factor analysis); the experimental manipulation of the construct (e.g., anxiety) and notation of change resulting in the test scores; and the study of differences in group performances on the test.

Leona Tyler gives two short definitions of construct validity:

> . . . A type of correlational research that does not involve any concrete empirical criteria of the traits under consideration. . . .
>
> A test has construct validity if the relationships between scores obtained on it and various other measures entering into the theoretical formulation turn out to be significant and in the predicted direction.
>
> (1965, p. 41)

Read the following example and analyze it in reference to these definitions.

A group, A, of people have been given a test on anxiety, and rated as "high anxious" or "low anxious."

A group, B, of theorists have developed an instrument measuring a new theoretical construct called "thritz." Group B has theorized that low-anxious people will have a low score on thritz, and high-anxious people will have a high score on thritz. This theory indicates that the ratings on group A will correlate very highly (near +1.00) with the thritz test.

Group A is given the thritz test, and the results are correlated with the anxiety ratings. The findings and next steps for several possibilities are given in Table 6.

TABLE 6 RESEARCH FINDINGS

Findings	*Next Step*
$r = .98$	The theory was correct for this group. Further testing of other groups is in order.
$r = .21$	Though there is some correlation, it is not sufficiently high for acceptance of the theory. Reevaluate the theory and/or revise the items.
$r = -.97$	The correlation shows that thritz correlates very highly but in the opposite direction from that theorized. *High-*anxious people are *low* on thritz. Validation studies are in order.

Keep in mind that one does not test the constructs, but rather the theory behind the constructs.

---

1. Abstract ideas or theories can be considered as ____________ which provide hypotheses that can be investigated.

---

constructs

---

2. Construct validity is therefore an indication of the relationship between a theory and ____________.

---

actual test performance

---

3. If a test does in fact measure the construct, the test has __________.

---

construct validity

---

4. How does one proceed with a test when the results of validation do not support the theory?

The constructs underlying the items or the theory may be reexamined.

---

5. How does one identify the constructs measured by a test when the main concern is with predictive validity?

To identify the constructs when determining predictive validity, one may use factor analysis to study the interrelationship of the items or manipulate the construct and note changes in the scores.

---

6. If one finds that his test has low construct validity, is there any rationale for knowing *why* this is the case so that improvements can be made? __________ Explain.

Yes. One can either take the position that the theory is inaccurate or that the items are measuring something else. Improvements can therefore proceed accordingly.

Face Validity

The primary types of validity which you will encounter are content, predictive, concurrent, and construct. However, there are several other types with which you should be familiar. For example, in certain circumstances, one may be concerned with the question: "Does the test *appear,* from examination of the items, to measure what one wishes to measure?" Or "Does the test *appear* to test what the name of the test implies?"

This is usually the concern of the layman who knows little or nothing about measurement, validity, or reliability. Public relations personnel often require "high face validity" for popular tests which they use in testing programs for industry, the military, and schools. However, it is difficult, if not impossible, to *measure* a validity of this type.

Face validity can also serve other functions. When one wishes to test for mental illness, it would be better to give a test with low face validity, as opposed to a test for creativeness or mechanical ability, which should have high face validity.

1. When one *examines,* but does not test, the relationship of the items on a test with its title, the primary concern is to establish the __________ validity of the test.

---

face

---

2. The layman is often concerned with face validity. He would like the test to have __________ face validity if he wants to convince the
(high/low)
user that he is testing what he says he is testing.

---

high

---

3. It is sometimes desirable to have low face validity for a test. A situation in which you might want to hide the true nature of the test from a user might be __________.

---

psychological, emotional, or pertaining to any undesirable trait

---

4. On the basis of the discussion in the text, would you assume that correlational techniques are used to establish the face validity of a test? ________________.

The authors hope not!

---

5. You should recall that one way of demonstrating the content validity of a test is to use a ________________ which shows topics covered by the test and types of items.

---

matrix or table

---

6. Discuss the rationale for using a content matrix to demonstrate the face validity of a test.

Since there is no statistical way to measure face validity, the use of a content matrix would offer an easy way to evaluate items from the test.

Curricular Validity

The term *curricular validity* was introduced by Cronbach (1960) with reference to the relationship between content of test items and objectives of instruction which has been completed. In an era of constantly changing curricula, the test producer finds he must revise his tests to maintain their curricular validity. A standardized achievement test which must have content validity seldom covers the exact objectives of any single course of instruction.

Curricular validity should not be confused with content validity. Content validity refers to the specific designation of the *topics* which are covered in the test and how well the test samples general situations or behaviors associated with those topics. Curricular validity refers to the observable relationship between the test and the instructional objectives of a specific curriculum.

1. The classroom teacher, who is concerned with the students' achievement of quite specific objectives, is therefore most concerned with the ________________ validity of a test.

---

curricular

---

2. Discuss the similarities and differences of content, face, and curricular validity.

All three are special types of validity. *Content* is concerned with how well the items in the test are related to a given area, *curricular* with how well given course objectives are represented in the test, and *face* with how well the test *appears* to measure an area.

Differential Validity

Differential validity refers to the difference between the correlations of a classification test with two separate criteria which are to be predicted (Anastasi, 1961, p. 182). The criteria might, for example, be success as a mechanic and success as an airplane pilot. If the classification test correlates .95 with success as a mechanic and .02 with success as an airplane pilot, the difference between the correlations would be .93 (that is, .95 − .02 = .93). The differential validity of the classification test is therefore considered to be .93 when the criteria of success as a mechanic and airplane pilot are employed. Thus differential validity indicates not only what a test *will* predict, but also what it *will not* predict.

Here is another example. Criteria: (1) success as a gardener, (2) success as a mechanic. Correlations:

	Gardener	*Mechanic*
Classification test	−.80	.95

The −.80 correlation indicates that persons who receive the highest scores on the classification test tend to do very well as a mechanic but very poorly as a gardener.

The differential validity would be

$$r_{t1} - r_{t2} = .95 - (-.80)$$
$$= 1.75$$

Most test batteries predict success or failure for a particular criterion, with no discrimination among various criteria. Differential validity is thus somewhat related to construct validity in that it facilitates the user's interpretation of the test scores.

It should be noted that differential-validity scores do not range from +1.00 to −1.00 as with a correlation coefficient, but may range from .00 to 2.00. (*Note:* Only absolute values are used; $.10 - .12 = -.02$, $.12 - .10 = +.02$.) Only the absolute value (.02) is used when reporting the differential validity of a test with a specified criterion. In the example above, in which the test correlates "high positive" with one criterion and "high negative" with another, the result will be greater than 1.

1. When the *difference* between two correlation coefficients is reported as a type of validity, it is a report on the ____________ ____________ of the test.

---

differential validity

---

2. The primary contribution of a differential-validity study is that it not only helps to indicate what the test does measure, but also ________.

---

what it does not measure

---

3. What is the differential validity of the following three tests?

	Criterion x	*Criterion* y
Test 1	.20	.90
Test 2	−.30	.70
Test 3	.60	.50

.70, 1.00, .10

---

4. Which of the tests in question 3 has the highest differential validity for the two criteria x and y? ________

---

test 2

---

5. How would you interpret the results of test 3 in terms of differential validity?

Test 3 shows low differential validity.

---

6. If you were not concerned about the differential validity of a test, how would you interpret the results of test 3?

Test 3 predicts both x and y about equally well.

From the foregoing discussion of the various types of validity, it should be obvious that one cannot say that "The *validity* of this test is. . . ." The manner in which it was obtained, the sample used, what criterion was utilized, and any other known restrictions must be stated in order to define "the validity of this test"—all this is in addition to stating whether it has concurrent, predictive, or some other type of validity.

Review Questions

1. List the types of validity discussed.
2. List the four types used most commonly.
3. Define "criterion."
4. Define "validation group."
5. What is "validity"?
6. Why must a test be valid?
7. How does one validate a test?
8. Does a test have only one validity? How many ways can a test be validated?
9. When the validity is reported at .92, what else must one know in order to be able to use the test efficiently?
10. What statistical procedure is generally used to determine a validity coefficient?

6

Statistical Procedures for Calculating Validity Coefficients

You have been introduced to a variety of types of validities and their uses. It is important that you have an understanding of the various statistical procedures (correlations) which are used for deriving validity coefficients.

Discussion of Pearson Product-Moment Coefficients

In the discussion of reliability, the following formula was used for computing the correlation between two sets of scores:

$$r = \frac{N\Sigma XY - \Sigma X \Sigma Y}{\sqrt{[N\Sigma X^2 - (\Sigma X)^2][N\Sigma Y^2 - (\Sigma Y)^2]}}$$

The r (correlation coefficient) may be written r_{xx} when the formula is used to determine reliability, since theoretically it is the correlation between test$_x$ and itself (test$_x$). The X and Y in the formula are both representative scores on test$_x$. Whether one obtains a reliability coefficient via the alternate forms, test-retest, or split-half methods, one obtains, essentially, a correlation between a test and itself.

From the discussion of validity it should be clear that the obtained correlation is not between a test and itself, but between test$_x$ and some external criterion$_y$. Then

r_{xy} = validity coefficient for test$_x$ with criterion$_y$

Therefore the previous formula may also be written for determining a validity coefficient:

$$r_{xy} = \frac{N\Sigma XY - \Sigma X \Sigma Y}{\sqrt{[N\Sigma X^2 - (\Sigma X)^2][N\Sigma Y^2 - (\Sigma Y)^2]}}$$

where X is the score on test$_x$, and Y is the score on criterion$_y$.

The correlation formula described above is a *product-moment correlation.* It is not the only one, even though the above formula is referred to as *the* Pearson *product-moment.* There are in fact three basic product-moment correlations, and two other related correlation coefficients, which will be discussed. Refer to Table 7 as you read the descriptions which follow. (For a review of the concept of correlation, consult Appendix C. A brief review is also given in the questions below.)

1. Indicate whether the five correlations in Fig. 3 are high or low or positive or negative or whether the correlation is about zero.

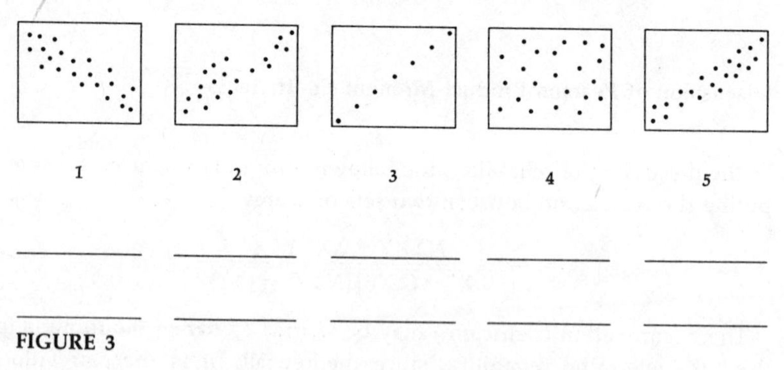

FIGURE 3

high negative, high positive, very high positive (+1.00), zero, high positive

2. Units on the vertical axis usually represent the criterion or the ______________ scores.
(x or y)

---

y

---

3. Units on the horizontal axis usually represent the test or the ______________ scores.
(x or y)

---

x

---

4. r_{xx} refers to a correlation between ______________.

---

a test and itself

---

5. r_{xy} refers to a correlation between test_x and ______________.

---

criterion_y

The *general product-moment correlation* is used when both the test scores and the criterion scores (or more generally, *both variables*) are continuous. A continuous distribution is one in which the outcomes are divided into equal-size units, such as time, temperature, grades, etc. When both the test results and the criterion results are measured on a continuous scale, and the scores on each approximate a normal distribution, then the appropriate validity coefficient is the product-moment correlation.

The *point-biserial correlation* is utilized when one variable is continuous and one variable is dichotomous. A *dichotomous* variable is one which can have only one of two possible values, typically, 1 or 0. Yes-no, male-female, pass-fail are variables of this type. The first of each pair may be assigned 1, the second 0, or vice versa, for purposes of calculation.

The *phi coefficient* must be employed when both variables under consideration are true dichotomies. For example, sex (male-female) might be used to predict success or failure on a particular task. However, the results of most tests and criteria which are encountered in educational and psychological testing rarely form a true dichotomy. More often than not, raw scores, percentages, or standardized scores are used. At times, though, there is good reason to want to *force* continuous scores into a dichotomy. When this technique is employed, it is necessary to use either the biserial or tetrachoric correlation coefficients.

TABLE 7 CORRELATION COEFFICIENTS

Product-Moment Coefficient	*Variable*		*Related to Product-Moment Coefficient*	*Variable*	
	X *Test*	Y *Criterion*		X *Test*	Y *Criterion*
General product-moment Pearson (r)	Continuous	Continuous			
Point-biserial (r_{pb})	Continuous	True dichotomy	Biserial (r_b)	Continuous	Continuous, but forced into a dichotomy
Phi (ϕ)	True dichotomy	True dichotomy	Tetrachoric (r_t)	Continuous, but forced into a dichotomy	Continuous, but forced into a dichotomy
Assumption: For true dichotomy, (1) dichotomous outcomes, (2) independent trials, (3) constant probability of success. Approximates normal population if $p = .5$, N is large.			*Assumption:* Data were normally distributed before dichotomizing them. Estimates correlation as if continuous data had been used.		

The *biserial correlation* is related to the point-biserial coefficient in that one variable (either the test or the criterion) is continuous, but the other is *not* a *true* dichotomy; the scores have been forced into a dichotomy. In this instance, the continuous variable might be scored pass-fail by scoring a "pass" for scores between 50 and 100 percent and "fail" for scores below 50 percent. Though the criterion scores are essentially continuous, they have been forced into a dichotomy.

The *tetrachoric coefficient* is used when both variables have been forced into dichotomies. As with the biserial, the assumptions underlying the use of the tetrachoric are that the correlation obtained is as it would have been if normally distributed continuous data had been used.

1. List five correlation methods:

1. Product-____________
2. Bi____________
3. P_ _ _ _ _ - ____________
4. P_ _
5. Tetra_ _ _ _ _ _ _

---

moment, biserial, point-biserial, phi, tetrachoric

---

2. Which coefficient is used when both the test scores and the criterion scores are continuous? ____________

Pearson product-moment

---

3. Which coefficient is used when one variable is on a continuous scale and one a true dichotomy? ____________

---

point-biserial

---

4. What is the assumption about the scores which are forced into a dichotomy?

The dichotomous scores are from a normally distributed, continuous distribution.

---

5. Complete as much of the following table as you can remember. Then refer to Table 7 to finish. Practice Table 7 until you reach perfection.

Coefficient	Variable X	Variable Y
Pearson r	________	________
Point-biserial	________	________
Phi	________	________
Biserial	________	________
Tetrachoric	________	________
Rho	Both X and Y can be ranked.	

Point-biserial Correlation

In the following sections you will be given hypothetical situations in which each of the various procedures for estimating validity will be demonstrated. Formulas will be explained and data analyzed. Each section will be completed with a brief discussion of the procedures. (*Note:* Carry all decimals to *three* places and round to *two*. *Example:* $8\overline{)3.000}$ = .375 = .38.)

Recall that the point-biserial correlation is used when one variable (either the test score or the criterion score) is continuous and the other is a natural dichotomy. The formula for the point-biserial correlation coefficient is

$$r_{\text{pbis}} = \frac{\overline{X}_p - \overline{X}_t}{\sigma_t}\sqrt{\frac{p}{q}} \qquad \text{Point biserial}$$

where

r_{pbis} = point-biserial correlation coefficient

$\overline{X}_p$ = mean score on the continuous variable (Y, in this case) for all examinees who were successful (or had a score of 1) on the dichotomous test (X, in this case)

$$\frac{\Sigma XY}{N_p} \qquad N_p = \Sigma X$$

$\overline{X}_t$ = mean score for all examinees on the continuous variable

σ_t = standard deviation of all scores on the continuous variable

p = proportion of all examinees classified as "1" on the dichotomous variable, or $\Sigma X/N$

$q = 1 - p$, or proportion of examinees with 0's on the dichotomous variable

An example of a situation in which the point-biserial coefficient would be appropriately used follows.

A researcher hypothesizes that the number of traffic tickets received by an individual can be predicted from his performance on his initial driver's test [which is scored simply pass (1) or fail (0)]. A sample of driving records resulted in the data in Table 8.

$$p = 10/15 = .67 \qquad \overline{X}_t = 50/15 = 3.33$$

$$\overline{X}_p = \frac{5+1+4+7+1+6+2+6+4+2}{10} = \frac{38}{10} = 3.80$$

$$\sigma_t = \sqrt{\frac{(5-3.33)^2 + (3-3.33)^2 + \cdots + (2-3.33)^2}{N}}$$

$$= \sqrt{232/15 - (3.33)^2}$$

$$= \sqrt{15.47 - 11.09} = 4.38 = 2.1$$

$$r_{\text{pbis}} = \frac{3.80 - 3.33}{2.1}\sqrt{\frac{.67}{.33}} = (.22)\sqrt{2.03} = (.22)(1.42)$$

$$= .31$$

DISCUSSION: One must be careful in interpreting r_{pbis} in terms of

TABLE 8 DRIVER'S TEST PERFORMANCE AND TRAFFIC TICKETS OBTAINED

Driver ID	X *Driver's Test*	Y *No. of Traffic Tickets*
01	1	5
02	0	3
03	1	1
04	1	4
05	1	7
06	0	0
07	1	1
08	1	6
09	0	5
10	1	2
11	1	6
12	0	3
13	0	1
14	1	4
15	1	2
	$\Sigma = 10$	$\Sigma = 50$

noting the way in which the dichotomous variable is coded. In the example used on p. 107, a score of 1 was assigned to those who passed the driver's test on the first attempt and a score of 0 was assigned to those who failed. The criterion variable *Y* was the number of traffic tickets the person had received. Therefore the correlation of .31 indicates that those persons who pass the test on their first attempt tend to have more traffic violations than those who fail on their first attempt. However, the correlation is not high enough to use it with much success.

1. Given:

$$r_{\text{pbis}} = \frac{\overline{X}_p - \overline{X}_t}{\sigma_t}\sqrt{\frac{p}{q}}$$

and the following item and total test-score information:

Student	X Question 3	Y Grade
1	0	20
2	1	24
3	1	25
4	0	19
5	1	22
$N = 5$	$= 3$	$= 110$

$\overline{X}_p =$ ____ $\overline{X}_t =$ ____ $\sigma_t = 2.3$ $p =$ ____ $q =$ ____

Determine the point-biserial correlation coefficient.

$r_{\text{pbis}} =$ ____

$\overline{X}_p = 23.67$, $\overline{X}_t = 22.00$, $p = .6$, $q = .4$, $r_{\text{pbis}} = \underline{.89}$

HELP:

$$\overline{X}_p = \frac{\Sigma XY}{N_p} = \frac{(0)(20) + (1)(24) + (1)(25) + (0)(19) + (1)(22)}{\Sigma X}$$

$$= \frac{0 + 24 + 25 + 0 + 22}{3}$$

$$= 71/3 = \underline{23.67}$$

$$\overline{X}_t = \frac{\Sigma Y}{N} = \frac{110}{5} = \underline{22.00}$$

$$p = \frac{\Sigma X}{N} = \frac{3}{5} = \underline{.6}$$

$$q = 1 - p = 1 - .6 = \underline{.4}$$

$$r_{\text{pbis}} = \frac{23.67 - 22.00}{2.3}\sqrt{\frac{.6}{.4}}$$

$$= \frac{1.67}{2.3}\sqrt{1.5} = (.73)(1.22) = \underline{.89}$$

---

2. On the basis of the point-biserial computed in question 1, what can you say about the validity of using test item 3 to predict a student's total score?

A teacher who had time for only one question in the assessment of the class could use question 3 with a fair degree of confidence.

---

3. Why would you use a point-biserial correlation for the type of data given in question 1?
(*Note:* the point-biserial is typically used in the item analysis of a test. In this context, it is referred to as the *discrimination index* for an item.)

One variable is continuous, and the other is a true dichotomy.

---

4. Given:

$$N = 10 \qquad N_p = 6$$
$$\Sigma XY = 120 \qquad \overline{X}_p = \_\_\_\_\_\_\_\_$$
$$\Sigma X = 200 \qquad \overline{X}_t = \_\_\_\_\_\_\_\_$$
$$\sigma_t = 10.0 \qquad r_{\text{pbis}} = \_\_\_\_\_\_\_\_$$
$$p = .6$$

$\overline{X}_p = \underline{20}$, $\overline{X}_t = \underline{20}$. With this information, examination of the formula will show that $r_{\text{pbis}} = \underline{0.}$

HELP:

$$\overline{X}_p = \frac{\Sigma XY}{N_p} = \frac{120}{6} = 20$$

$$\overline{X}_t = \frac{\Sigma X}{N} = \frac{200}{10} = 20$$

$$r_{\text{pbis}} = \frac{\overline{X}_p - \overline{X}_t}{\sigma_t}\sqrt{\frac{p}{q}}$$

$$= \frac{20 - 20}{10}\sqrt{\frac{.6}{.4}}$$

$$= 0$$

Phi Correlation

If a researcher were interested in trying to predict who is likely to pass a driver's test, he might use such variables as graduation or nongraduation from high school, or sex (male or female). In this situation, in which both variables form a natural dichotomy, the phi coefficient is used.

The phi correlation coefficient formula is

$$\phi = \frac{p_{ij} - p_i p_j}{\sqrt{(p_i - p_i^2)(p_j - p_j^2)}} \qquad \text{Phi}$$

where

p_{ij} = proportion of examinees who perform successfully on both variables

p_i = proportion of examinees who perform successfully on predictor variable

p_j = proportion of examinees who perform successfully on criterion variable

or the formula may be written

$$\phi = \frac{p_{ij} - p_i p_j}{\sqrt{(p_i q_i)(p_j q_j)}}$$

The researcher might choose to investigate the relationship between sex and success on the driver's test. Table 9 shows a set of hypothetical data he might obtain.

The phi coefficient would be determined as follows:

$p_i = 5/10 = .50 \qquad q_i = .50$

$p_j = 4/10 = .40 \qquad q_j = .60$

$p_{ij} = 3/10 = .30$

$$\phi = \frac{.30 - (.50)(.40)}{\sqrt{[(.50)(.50)][(.40)(.60)]}}$$

$$= \frac{.30 - .20}{\sqrt{(.250)(.240)}}$$

$$= \frac{.10}{\sqrt{.060}} = \frac{.10}{.24} = .42$$

TABLE 9 DRIVER'S-TEST RESULTS AND SEX OF THE INDIVIDUALS

	Variable									
Sex*	1	0	1	1	0	0	0	0	1	1
Driver's test†	0	0	0	1	0	0	1	0	1	1

*Male = 1, female = 0, (*i*).
†Pass = 1, fail = 0, (*j*).

$$q_i = 1 - p_i = 1 - .50 = .50$$

$$q_j = 1 - p_j = 1 - .40 = .60$$

DISCUSSION: Based on the hypothetical data in the sample, the validity coefficient of predicting whether a person passes his driver's test from the sex of the individual is .42. If the hypothetical validating sample were representative of those individuals who apply for the driver's test, one could argue that there is some validity in predicting their success based on their sex. However, the coefficient is not sufficiently high to be convincing.

This is a good sample problem in which to indicate that a validity coefficient of −.42 is equally as "good" as one which is +.42. The direction of the correlation, either positive or negative, is dependent upon the rather arbitrary assignment of 1 and 0 to the two possible states of the dichotomous variable. In this example, if females had been coded 1 and males 0, the resulting phi coefficient would have been −.42.

When one uses ϕ, the data are binomial; that is, all the values obtained in the raw data are 1's or 0's. Therefore proportions are more sensible to use than any formula that requires a ΣX^2, since the ΣX^2 for a ϕ problem would be the same as ΣX. An easy way to remember this relationship is: *p*hi uses *p*roportions.

1. Given:

0 = yes
1 = no

Survey Questionnaire										
Question 1	0	0	1	0	1	1	1	0	1	1
Question 2	0	1	1	1	0	1	0	0	1	1

QUESTION: If a person answers the table question 1 "yes," can you predict from observing the results shown above that he will also answer question 2 "yes"? ____________

No, there appears to be a low correlation between 1 and 2

---

2. What correlation coefficient would be appropriately used with the data in question 1, and why? ____________

Phi, or ϕ. Both parts of question 1 are true dichotomies.

---

3. Estimate the phi coefficient for the data in question 1. The formula for phi is

$$\phi = \frac{p_{ij} - p_i p_j}{\sqrt{(p_i q_i)(p_j q_j)}} = \_\_\_\_\_\_\_\_$$

$$\frac{.40 - .36}{\sqrt{(.6)(.4)(.6)(.4)}} = \frac{.04}{.24} = .17$$

HELP:

$$\sqrt{(p_i q_i)(p_j q_j)} = \sqrt{(.6)(.4)(.6)(.4)} = \sqrt{(.24)(.24)} = .24$$

---

4. When is the phi coefficient used? ______________

____

when both variables are dichotomous

____

Biserial Correlation

In essay exams it is difficult to grade on a very accurate scale. However, if one had to assign a grade of pass or fail, he would probably do it with some scale in mind. This student deserves a "high pass," that one a "low pass," another "average." Pass-fail is, in this case, a forced dichotomy based on continuous data which range from "high pass" to "fail." Another example of the use of a forced dichotomy is the arbitrary decision that all grades of 60 percent and above pass a test and all those below 60 percent are failures.

If the essay (pass-fail) test is used as a criterion measure for a multiple-choice test where possible scores range from 0 to 100, the biserial correlation is used. A typical example of this situation would be one in which the researcher would like to establish the concurrent validity of a multiple-

TABLE 10 HYPOTHETICAL DATA FOR TWO SETS

Student	X *Multiple-choice-test Score*	Y *Essay Score**
1	25	0
2	30	1
3	20	1
4	25	0
5	40	1
6	30	0
7	20	1
8	35	0
9	25	1
10	45	1

*0 = fail; 1 = pass.

choice test which purports to measure essay-writing skill. He might use the consensus of judges' ratings of whether actual essays written by the students were of "passing" or "failing" quality.

Hypothetical data are given in Table 10.

The formula for the biserial correlation is

$$r_{\text{bis}} = \frac{\overline{X}_p - \overline{X}_t}{\sigma_t}\left(\frac{p}{z}\right) \qquad \text{Biserial}$$

where r_{bis} is the biserial correlation coefficient, and z is the value in the unit normal distribution corresponding to p. All other symbols are the same as those in the r_{pbis} formula.

Most statistical textbooks have tables which indicate the z, or "area in the larger portion" of the distribution, and ordinates for areas under the normal curve. Sample values are as follows:

Area in Larger Portion (p *or* $1-p$)	*Ordinate* z
.50	.399
.60	.386
.70	.348
.80	.280
.90	.175
.95	.103
.99	.027

The value for z which is used in the biserial formula is directly related to the value of p (the proportion of examinees who are successful on the dichotomous variable). Therefore, if p is .80, the corresponding value of z is .280. Since the normal distribution curve is symmetrical, it is possible to determine the z for values of p which are less than .50 by subtracting the p value from 1. For example, if p equals .30, then $1 - p$ equals .70. Using this value for entering the table, we find that the appropriate z value is .348.

For the data shown on Table 10, the formula would be used as follows:

$$N_p = \Sigma Y = 6 \qquad N = 10$$

$$\overline{X}_p = \frac{30 + 20 + 40 + 20 + 25 + 45}{N_p}$$

$$= 180/6 = \underline{30}$$

$$X_t = \frac{25 + 30 + 20 + \cdots + 35 + 25 + 25}{N}$$

$$= 295/10$$

$$= \underline{29.5}$$

$$\sigma_t = \sqrt{\frac{(25 - 29.5)^2 + (30 - 29.5)^2 + \cdots + (45 - 29.5)^2}{N}}$$

$$= \sqrt{\frac{(4.5)^2 + (.5)^2 + \cdots + (15.5)^2}{10}}$$

$$= \sqrt{\frac{661.5}{10}} = \sqrt{66.15} = \underline{8.12}$$

$$p = \tfrac{6}{10} = .6$$

$$z = \underline{.386}$$

$$r_{\text{bis}} = \frac{30 - 29.5}{8.12} \times \frac{.6}{.386}$$

$$= \left(\frac{.5}{8.12}\right)(1.55)$$

$$= (.06)(1.55)$$

$$= .09$$

DISCUSSION: How is the biserial r to be interpreted? It is like any other correlation, in that it indicates the amount of variance in the scores on one variable which can be predicted from another variable. However, it should be noted that, for statistical reasons, the biserial coefficient can exceed the range from minus 1 to plus 1.

The concurrent validity coefficient for using the multiple-choice-test score to predict a student's essay grade in this situation is .09. With a coefficient this low, it is clear that the ability that is measured by the multiple-choice test is not related to the ratings of the students' essays.

The validity coefficient for Test X is not the only one which could be obtained. If the test were given to another group of individuals under a different set of circumstances (such as changing the directions to the judges), another validity coefficient could be obtained.

1. When does one use a biserial correlation? ______________

When one variable is continuous and one is dichotomized.

2. Given:

$$r_{\text{bis}} = \frac{\overline{X}_p - \overline{X}_t}{\sigma_t} \frac{p}{z}$$

$$\overline{X}_p = 8.5 \qquad \overline{X}_t = 10$$
$$\sigma_t = 2.0$$
$$p = .50$$
$$z = .399$$

What is the biserial coefficient?

$r_{\text{bis}} =$ ______________

$$\left(\frac{8.5 - 10.0}{2.0}\right)\left(\frac{.50}{.399}\right) = -.938$$

3. If the standard deviation of the total scores changes from 2.0 to 3.0, does the biserial coefficient increase or decrease in question 2? Work the problem to check your answer. ______________

Decrease, $\left(\frac{8.5 - 10.0}{3.0}\right)\left(\frac{.50}{.399}\right) = -.625$

4. Given:

$$\overline{X}_p = 80.5$$
$$\overline{X}_t = 65.0$$
$$\sigma_t = 20.0$$
$$p = .40$$
$$z = .386$$

Find:

$$r_{\text{bis}} = \underline{\hspace{6em}}$$

$$\left(\frac{80.5 - 65.0}{20.0}\right)\left(\frac{.40}{.386}\right) = \underline{.80}$$

---

5. Find the r_{bis} and discuss the correlation obtained.

Given:

Multiple-choice, X	Essay, Y
95	0
90	0
60	1
95	0
60	1
50	1
40	1
65	1
50	1
85	0

Find:

$N =$ ____
$\overline{X}_p =$ ____
$\overline{X}_t =$ ____
$\sigma_t =$ ____
$p =$ ____
$z =$ ____
$r_{\text{bis}} =$ ____

$$\overline{X}_p = \frac{(X)(Y)}{Y}$$

$$\sigma_t = \sqrt{\frac{(X-\overline{X})^2}{N}}$$

$N = 10 \quad \overline{X}_p = 54.17 \quad \overline{X}_t = 69.00 \quad \sigma_t = 20.54$

$p = .6 \quad z = .386 \quad r_{\text{bis}} = -1.12$

High negative correlation > 1.00. The lower the grade on the essay, the higher the grade on the multiple-choice test.

HELP:

$$\overline{X} = \frac{60 + 60 + 50 + 40 + 65 + 50}{\Sigma Y}$$

$$= 325/6 = 54.17$$

$$\sigma_t = \sqrt{\frac{(95-69)^2 + (90-69)^2 + \cdots + (50-69)^2 + (85-69)^2}{N}}$$

$$= \sqrt{3{,}790/10} = \sqrt{379.0} = 19.5$$

$$r_{\text{bis}} = \left(\frac{54.17 - 69.00}{19.5}\right)\left(\frac{.6}{.386}\right) = \left(\frac{-14.83}{19.5}\right)(1.55)$$

$$= (-.760)(1.55) = -1.178$$

ΣY in this case represents N_p, or the number of those who received a 1 on the Y variable.

It should be noted that, having obtained a biserial coefficient, a relatively simple method is available for obtaining the point-biserial coefficient. The formula is

$$r_{\text{pbis}} = r_{\text{bis}} \frac{z}{\sqrt{pq}}$$

In the sample problem associated with the biserial coefficient, $r_{\text{bis}} = .09$, $z = .386$, $p = .6$, and $q = .4$. Therefore, for the same data,

$$\begin{aligned} r_{\text{pbis}} &= .09 \frac{.386}{\sqrt{(.6)(.4)}} \\ &= (.09)\left(\frac{.386}{.49}\right) \\ &= (.09)(.79) \\ &= .07 \end{aligned}$$

In all cases, the point-biserial equivalent of the biserial coefficient will be lower; i.e., it is a more conservative estimate.

Tetrachoric

When both variables are forced into dichotomies, the tetrachoric correlation is required. The tetrachoric formula estimates the correlation between the two variables as if they were both continuous.

Theoretically, every variable which is typically used in psychological measurement is a continuous one. Therefore any dichotomy is necessarily forced. A simple yes-no dichotomy is forced upon people who would answer wholeheartedly or with reservations. An example could be, "Do you smoke?"

ANSWER: *Chain Smoker*   *Often*   *Sometimes*   *Seldom*   *Never*

Where would you place yourself? Would you give an unqualified "Yes" or "No" to the question? To dichotomize this variable, a decision line must be drawn somewhere. Would you draw it between "Seldom" and "Never" or between "Sometimes" and "Seldom"?

Some test analysts would argue that all analysis of test data should use the tetrachoric for this reason. However, because of its complicated procedures, the tetrachoric is not often used.

The tetrachoric correlation is the most laborious of the correlation coefficients to calculate. It is based on complicated mathematics which at this point you are not expected to know. Let it suffice that you are aware it exists and under what circumstances it is to be used. A simple example will give you a rough idea of the method.

Suppose an aptitude test is scored from 0 to 100 but the students are separated into pass-fail on the basis of a cutting point of 50. All students above 50 pass, all with 50 or below fail.

This same group of students was admitted or rejected from the state college on the basis of a 2.5 high school grade point average. The variable of grade point average is a continuous variable from .0 to 4.0, which had been dichotomized in order to facilitate selection procedures. The question can be asked whether the aptitude test has any predictive validity in terms of college admissions.

The data are presented in Table 11, in a fourfold table.

TABLE 11 FOURFOLD PRESENTATION OF TWO VARIABLES

Aptitude Test Scores	*Admission to College*		
	No	*Yes*	*Total*
0–50	20 (*a*)	30 (*b*)	50
51–100	10 (*c*)	40 (*d*)	50
Total	30	70	100

$N = 100$

Using the cell frequencies from the matrix, it is now possible to use the short-cut formula ad/bc, to estimate the tetrachoric correlation.

$$\frac{(a)(d)}{(b)(c)} = \frac{(20)(40)}{(30)(10)} = \frac{800}{300} = 2.67$$

Entering a table for ad/bc values (which may be found in many statistical texts), an r_{tet} of .36 is obtained. The aptitude test is not a good predictor of college admission, with a validity of .36. Note that this is an estimation of the tetrachoric correlation. The tetrachoric correlation is less reliable than the Pearson *r;* for example, it requires approximately twice as many cases with a tetrachoric correlation to achieve a coefficient as reliable as that obtained by the product-moment method.

Note that the tetrachoric formula *ad/bc* must always produce a value greater than 1. For this reason, it is immaterial whether you label the matrix

	Yes	No	Total
Yes	a	b	$a+b$
No	c	d	$d+c$
Total	$a+c$	$b+d$	$a+b+c+d$

or

	Yes	No	Total
Yes	b	a	$a+b$
No	d	c	$d+c$
Total	$d+b$	$a+c$	$a+b+c+d$

The larger product is always placed in the numerator, the smaller product in the denominator position. Therefore

$$\frac{ad}{bc} = \frac{(30)(20)}{(40)(20)} = \frac{600}{800}$$

would be reversed to read

$$\frac{800}{600} = 1.33$$

From the *ad/bc* table for an estimation of r_{tet}, 1.33 gives an r_{tet} of .11.

•

1. When both variables are forced into dichotomies, validity is determined through the use of a ____________ ____________.

---

tetrachoric correlation

---

2. To compute an estimate of the tetrachoric coefficient, the frequencies of diagonal cells are multiplied and then divided. Which of the following represents this relationship?

____ a. $\frac{ab}{bc}$ ____ b. $\frac{ad}{bc}$

____ c. $\frac{ac}{bd}$ ____ d. $\frac{ac}{db}$

---

b

---

3. The correlation obtained by using the formula in question 2 is only a(n) ____________ of the tetrachoric.

---

estimate or approximation

---

4. The tetrachoric correlation is ____________ reliable than the
(*less/more*)

Pearson *r*.

---

less

---

5. Using the information below, estimate the tetrachoric correlation. Some estimates for r_{tet} for various values of ad/bc are provided.

		Q-1		
		Yes	No	Total
Q-2	Yes	30	20	50
	No	10	40	50
	Total	40	60	100

$ad/bc =$ ____

ad/bc	r_{tet}
1.67	.20
2.5	.35
4.4	.53
6.0	.62

$r_{tet} =$ ____

$$\frac{ad}{bc} = \frac{20 \times 10}{30 \times 40} = \frac{200}{1{,}200} \quad \text{reverse} \quad \frac{1{,}200}{200} = 6.0 \qquad r_{tet} = .62$$

Rank-order Correlation (ρ, or Rho)

This correlation is typically used when there are only a few cases (or subjects) involved and the data are ranked or can be ranked. There is no assumption that ranks are based on equal-interval measures, or that there is *any* measurement involved. The ranks are merely an indication that this person is first in this attribute or test, this person second, and so on.

The Spearman rank-order correlation-coefficient (rho) formula is

$$\rho = 1 - \frac{6\Sigma D^2}{N(N^2 - 1)}$$ Spearman rank-order correlation formula

where D is the difference between two ranked sets of scores, and N is the number of pairs of observations. If a student is ranked fourth on one variable and seventh on another, then $D = 3$ and $D^2 = 9$.

As an example, one might hypothesize that there would be a relationship between the order (or sequence) in which students signed up for a psychological experiment and their scores on an experimental task. Table 12 shows a hypothetical set of scores, designed to test this hypothesis.

TABLE 12 RANK-ORDER-CORRELATION EXAMPLE

(a) *Sign-up Order*	*(b)* *Score on Experimental Task*	*(c)* *Rank Order on Experimental Task*	*(d)* *D* *(c − a)*	*(e)* *D^2*
1	20	2	1	1
2	21	3	1	1
3	23	4	1	1
4	25	5	1	1
5	29	8	3	9
6	28	7	1	1
7	30	9	2	4
8	32	10	2	4
9	27	6	3	9
10	19	1	9	81
				$\Sigma = 112$

The order in which the students register for the experiment (column *a*) serves as one set of rankings. The test scores need to be ranked, however. The *lowest* score will be assigned the rank of 1, the next lowest a score of 2, etc., as has been done in column *c* of the data.

Column *d* is the difference between the rankings in column *a* and *c* irrespective of + or − signs, since squaring them in column *e* eliminates these.

Using the sum of column *e*, the rank-order correlation can be determined as follows:

$$\rho = 1 - \frac{(6)(112)}{(10)(10^2 - 1)}$$

$$= 1 - \frac{672}{990}$$

$$= 1 - .68$$

$$= .32$$

DISCUSSION: There is some correlation between the fact that persons who sign up first get the lowest scores. However, it is not always the case; so predictions made on that basis should be made with care, if at all.

Correlation may be computed between two variables which are in no way logically related or, if they are related, are not meaningful. A teacher might, for example, feel that it would be interesting to find the correlation between grades on a test and the order in which the tests were turned in. Interesting, yes, but not very useful.

1. When there is a relatively small number of scores, and these scores can be ranked, one can use the ______________-______________ ______________.

---

rank-order correlation, or rho (ρ)

---

2. When using the formula, one needs to find only two missing numbers. These are represented by the letters ____________ and ____________.

D and *N*

---

3. *D* stands for the difference between ____________ ____________.

---

ranked scores

---

4. *N* stands for the number of ______________ of scores.

---

pairs

---

5. Given:

$$\rho = 1 - \frac{6\Sigma D^2}{N(N^2 - 1)}$$

Find rho (ρ) for the following:

Score A	*Score* B	*Rank* *A*	*Rank* *B*	*D*	*D²*
12	6				
10	9				
2	8				
7	2				
40	12				
8	5				
					$\Sigma =$ ____

A	*B*	*D*	*D²*
5	3	2	4
4	5	1	1
1	4	3	9
2	1	1	1
6	6	0	0
3	2	1	1
			$\Sigma = \underline{16}$

$$\rho = 1 - \frac{(6)(16)}{(6)(36 - 1)}$$

$= 1 - 96/210$
$= .543$

7

Procedures and Considerations

This chapter gives special attention to the topics of cross-validation and attenuation. It considers the need for substantiating the results of a single study by administering the test instrument to a number of representative groups. It takes up, also, several techniques for estimating the "true" validity of an instrument by correcting for errors of measurement in the test scores, the criterion scores, or both. The effect of selective sampling is also noted.

Cross-validation

Cross-validation refers to the technique of confirming a validity coefficient by administering the test to another group and correlating that test with the criterion scores. Using another sample tends to substantiate the original results. A test whose items were specifically selected on the basis of the results obtained on a particular sample would have an even higher correlation if it were again given to *that group*. To give the test to *another sample* tends to eliminate chance factors. The resulting correlation is usually a lower estimate of the effectiveness of the test.

Only after the statistical analysis of a test has shown it to be effective does one conduct a cross-validation study. The test at this point should not be revised or changed in any way, or the resulting study will be a "further" validation, not a cross-validation. The scoring system used with the validating group must also be the same.

The group used for the cross-validation study must be similar to, but not identical with, the original group. If the cross-validation study results in essentially the same validity coefficient, the test can be made available for further research and limited practical application. If the results are substantially lower, more research is required.

1. When the validity of a test is checked using a new sample, it is said to be ______________ validated.

---

cross-

---

2. The validity coefficient of a test is usually ______________ when *(raised/lowered)* cross-validated.

---

lowered

---

3. Cross-validation tends to eliminate ______________ factors in initial correlations.

---

chance

---

4. A researcher carries out the following procedures:

1. Designs test
2. Administers test to group A
3. Correlates results with a criterion
4. Selects test items with highest point-biserial correlations
5. Reconstitutes test
6. Administers test to group B
7. Correlates test results with a criterion

Is this an appropriate cross-validation set of procedures? ____________

no

What steps should be eliminated to make it appropriate? ____________

steps 4 and 5

What steps could be added to make it appropriate? ____________

Replicate steps 6 and 7 using group C.

Correction for Attenuation

It has been repeatedly pointed out that criterion scores are often quite unreliable. Obviously, the test for which the validity is desired is also not perfectly reliable. Therefore the correlation coefficient between the two variables is lower than would be the case if both were perfectly reliable.

When two nonreliable measures (i.e., when r_{xx} and r_{yy} are less than 1) are correlated, the correlation coefficient is said to be attenuated. Attenuation means that the correlation is reduced because of uncorrelated errors of measurement of the imperfect variables. To correct for errors of measurement in both of the variables, the following formula is used:

$$r_c = \frac{r_{xy}}{\sqrt{r_{xx} r_{yy}}}$$ Correction for attenuation in test and criterion

where

r_c = corrected validity coefficient
r_{xx} and r_{yy} = reliability coefficients of two tests
r_{xy} = correlation between two tests

The correction procedure is utilized as follows: If test x has a reliability of .49, and criterion y has a reliability of .64, and the correlation between x and y is .40, the corrected validity coefficient will be

$$r_{c_{xy}} = \frac{.40}{\sqrt{(.49)(.64)}} = \frac{.40}{(.70)(.80)} = \frac{.40}{.56} = .71$$

The validity coefficient is increased from .40 to .71 after correction for attenuation. The actual validity of the test has not been changed. Validity is a theoretical construct, and the correction as indicated by the formula is measured only in theoretical terms. To say that the validity increases after correction for attenuation is to imply that the test has a fixed "validity" to begin with, and by changing the coefficient, one can change the validity. On the other hand, it is valuable to know how much the validity could be increased if both variables were perfectly reliable.

In reality, the test developer can often only increase the reliability of his test instrument, and therefore it is of interest to know the validity of a test if there were a perfectly reliable criterion. If it is possible to increase the coefficient to a large degree, the test constructor may consider his time well spent if he can increase the reliability of the test.

The formula for correcting the correlation due to measurement errors in the criterion *only* is

$$r_{cc} = \frac{r_{xy}}{\sqrt{r_{yy}}}$$ Correction for attenuation in criterion.

Thus, with a criterion which has a reliability of .64, as before, and a correlation with the test of .40, a correction for attenuation due to error in the criterion produces

$$r_{cc} = \frac{.40}{\sqrt{.64}} = \frac{.40}{.80} = .50$$

This results in a higher validity estimate than the original coefficient (.40), but one that is lower than when both variables are corrected for attenuation (.71).

1. Correction for attenuation is used to correct for ______________ variance in the scores.

---

error

---

2. Correction for attenuation ______________ predictive validity
(increases/decreases)

estimates.

---

increases

---

3. Test$_x$ has a reliability estimate of .64. Criterion test$_y$ has a reliability estimate of .25.

1. Correct for attenuation on both tests when they have a validity coefficient of .32. $r_c =$ ______________

$$r_{c_{xy}} = \frac{.32}{\sqrt{(.64)(.25)}}$$

$$r_c = .80$$

HELP:

$$r_c = \frac{r_{xy}}{\sqrt{r_{xx} r_{yy}}} = \frac{.32}{\sqrt{(.64)(.25)}} = \frac{.32}{(.8)(.5)} = \frac{.32}{.40} = .80$$

---

2. Correct for attenuation on the criterion only for the same validity coefficient.

$$r_{cc} = \frac{r_{xy}}{\sqrt{r_{yy}}} \qquad r_{cc} = \_\_\_\_\_\_\_\_$$

$$r_{cc} = \frac{.32}{\sqrt{.25}} = .64$$

---

3. Correct the following validity coefficients for attenuation:

r_{xx}	r_{yy}	r_{xy}	r_c	r_{cc}
.80	.70	.53	___	___
.92	.65	.71	___	___
.25	.36	.27	___	___
.49	.56	.30	___	___

$r_c = .70, .91, .90, .57$; $r_{cc} = .63, .87, .45, .40$

Restriction of Range

If one uses a selected subgroup to test the reliability and validity of tests and criterion, one changes the outcomes of both. When the ability range is restricted, the variance, and therefore the validity, are also restricted. This phenomenon has already been noted for reliability. The scores of a selected group of students of between 95 and 100 percent on a test will have a lower correlation with a criterion than will those of the entire group receiving between 1 and 100 percent. There is little variance in the restricted group with a range in scores from 95 to 100, or 5 points, whereas the entire group has a 99-point range ($100-1=99$) and a large variance. An example is shown in Table 13.

TABLE 13 TEST SCORES ON TEST$_x$ AND CRITERION$_y$

	Total Group							*Selected Group*			*Total* TG	*Total* SG
Test$_x$	1	4	8	11	16	26	25	16	26	25	91	67
Criterion$_y$	5	6	7	8	9	10	11	9	10	11	56	30

Test	*Criterion*	*Test*	*Criterion*
$N=7$	$N=7$	$N=3$	$N=3$
$\overline{X}=13$	$\overline{X}=8$	$\overline{X}=22.3$	$\overline{X}=10$
$\sigma^2=82.3$	$\sigma^2=4$	$\sigma^2=20.22$	$\sigma^2=.667$
$\sigma=9.1$	$\sigma=2$	$\sigma=4.5$	$\sigma=.82$

Table 14 gives a clearer view of the difference encountered when a total group is used and when the group is restricted to one portion of the score range.

Test users sometimes encounter this problem when they try to use a test with a selected group and find the validity coefficient lower than the original validity coefficient. It is not that the test is less valid, but that the group has been selected to give less variance. The reduction in variance results in a lower validity coefficient.

TABLE 14 VARIANCE AND VALIDITY FOR TWO GROUPS

	Test σ^2	*Criterion* σ^2	*Validity*
Total group $N = 7$	82.3	4.0	.92
Selected group $N = 3$	20.2	.7	.85

1. Restricting the ability range of a group ________________
(lowers/raises/does not change)
the validity coefficients on tests taken by that group.

---

lowers

---

2. It follows from question 1 that if one wants to *inflate* a validity coefficient artificially, he should use a test group which ________________.

---

is not restricted in ability or exceeds the intended ability range for the test

---

3. Restricting the ability range which is tested also restricts the ________________, thus lowering the ________________.

---

variance, validity

Relation of Validity to Reliability

The validity of a test cannot exceed the square root of the reliability of the test. A test which correlates .81 with itself (r_{xx}) cannot have a validity coefficient of more than $\sqrt{.81} = .90$ with any criterion. This is termed the *intrinsic validity index,* or the *index of reliability.*

$$r_{xy} = \sqrt{r_{xx}}$$ Index of reliability

The more important aspect of the relationship between reliability and validity is that, in order to maximize one, you may be required to reduce tne other. For example, you know that internal-consistency reliability formulas (for example, K-R 20) take into account the interitem correlations: the higher these correlations, the higher the reliability of the test. Therefore, when one is developing a test, it is natural to retain those items which correlate highly with each other and with the total score. This has the effect of making the test more homogeneous and more reliable.

However, it is almost always the case that the criterion behavior, which is to be correlated with the test scores, is not a homogeneous, single-factor behavior. Scores on the criterion may be a composite of a great many skills, and therefore, to measure these skills, one needs to retain such items in the predictor test. By retaining such items, the reliability will undoubtedly be reduced, but in a trade-off situation such as this, it is usually preferable to maximize validity at the cost of reduced reliability.

1. The validity of a test cannot exceed the square root of the reliability of that test. If the reliability of a test is .64, it cannot have a validity coefficient greater than ________________.

.80

---

2. If a validity coefficient cannot exceed .80, it means that the scores on a test with $r_{tt} = .64$ will not ______________ higher than .80 with any other set of scores.

---

correlate

---

3. Inspection of reliability formulas indicates that the greater the interitem correlation, the ______________ the reliability.

---

higher or greater

---

4. By selecting the "good" items for use in a revised version of a test, one is actually increasing the interitem correlations, thus increasing the ______________ and ______________ of the test.

---

homogeneity, reliability

---

5. Increasing the reliability of the test may tend to ______________ the validity of the test.

---

reduce

---

6. The validity of a test can be reduced if its content is made more homogeneous because the criterion is ______________.

---

usually heterogeneous, or not homogeneous

8

Reporting Test Validities

A test author is under several obligations when offering his test for use by others. With respect to validity, he should report *how* the test was validated, with what group(s), when the test was validated, and what type of validity was used. The American Psychology Association Standards for Educational and Psychological Tests and Manuals (1966) lists these as essentials for reporting on any test.

Statistical manipulations should also be reported, including confidence limits, or the probability of misclassification of an individual. The criterion should be described thoroughly, especially in terms of adequacy and relevance and irrelevance. When a test manual recommends that the test be used with a certain type of population, the validation group must be a comparable sample which is adequately described.

It is strongly recommended that you read the APA Standards for a more complete examination of the information required for adequate reporting of test statistics.

Let it also be acknowledged that the use of tests in any situation necessitates additional local validation studies, since each situation is unique, no matter how closely the groups and situations compare. A counselor or administrator in a school system or college should conduct validation studies with the students under his jurisdiction for all tests that will be of any consequence to the students. This enables him to make better predictions for future groups, who will more closely resemble the present group than a validation group located elsewhere.

1. What is the name of the manual used as a guideline for reporting test statistics? ________________

APA Standards

---

2. What are some of the considerations involved in reporting test data?

How the test was validated, type of validity, any corrections used, and validation group.

---

3. Why is it necessary to have a manual of this type?

To provide guidelines for use of tests throughout the profession.

With this study you have successfully completed the descriptions of reliability and validity. While it may not have equipped you to be an expert in these areas, you should be adequately prepared to review information provided by test publishers and to compute your own local statistics. It should lay the groundwork for beginning the study of the topics of reliability and validity in depth.

APPENDIX A

Symbols

D, d = difference between scores for one individual

K = number of items

K-R 20, K-R 21 = Kuder-Richardson formulas 20 and 21 reliability coefficients

N = number in total group

N_p = number of those in total group who received 1 on the dichotomous variable; also ΣX for dichotomous variables

P = proportion, or $\Sigma X/N$ for dichotomous variables

$\overline{p}$ = mean of p

p_i, p_j = proportion of those in total group who received 1 on either X_i or X_j variable

p_{ij} = proportion of those in total group who received 1 on both dichotomous variables; also $(\Sigma XY)/N$

ϕ = phi correlation coefficient

$q = 1 - p$

r = correlation coefficient

r_{xx} = reliability; also r_{tt}

r_{xy} = validity

r_c = correction for attenuation of correlation in criterion only

r_{cc} = correction for attenuation of correlation in criterion and test
r_{bis} = biserial correlation
r_{pbis} = point-biserial correlation
r_t = tetrachoric correlation
ρ = rho = rank-order correlation
SB_r = Spearman-Brown correction formula
SE_m = standard error of measurement
σ = sigma = standard deviation; σ_t of total group
$\sigma_e{}^2$ = error variance
$\sigma_o{}^2$ = obtained or actual variance
σ_t^2 = true variance
X, Y = variables
X_e = error score
X_o = obtained score
X_t = true score
$\overline{X}, \overline{Y}$ = means of variables; $(\Sigma X)/N$, $(\Sigma Y)/N$
$\overline{X}_p$ = mean of scores on X with corresponding score of 1 on dichotomous Y variable; $(\Sigma XY)/N_p$
$\overline{X_t}$ = mean of total scores; also $\overline{X}$; $(\Sigma X)/N$
Σ = sum
ΣX = total of all X scores
$\sum_{i=1}^{N}$ = sum of i from 1 to N

APPENDIX B

Square Root

Examples of Square-root Problems

Listed below are the solutions to three square-root problems. The purpose of these problems is to refresh your memory of the process used to derive a square root. If these sample problems do not provide sufficient information, refer to any basic mathematics text for a detailed description of the process of extracting square roots.

```
            2 8 . 9  3   = 28.9
          √8 37 . 00 00
            4
            4 37
     48_    3 84
              53 00
     569_     51 21
                1 7900
     5783_      1 7349
                   551
```

```
      5 . 6  9  = 5.7              . 8  2  = .8
    √32 . 48 32                   √. 67 32
      25                              64

      7 48                            3 32
106_  6 36                     162_   3 24
      1 1232                             8
1129_ 1 0161
        1071
```

APPENDIX C
Correlation

If you have never formally studied the concept of correlation, you should refer to any standard statistics text for a thorough reading of the chapter dealing with this topic. The concept of correlation is crucial to the understanding of much of what has been written about reliability and validity, for it is the correlation coefficient which indicates the extent to which two things are related, or the extent to which a change in one thing is related to the change in another.

A common example of the use of correlation coefficient to indicate the degree of relationship between two things would be the relating of scores of a group of students on standardized verbal and quantitative aptitude tests. Note that there are two sets of scores for the same group of people. This is always a requirement when computing a correlation coefficient. The scores on the two tests would be used in the product-moment correlation formula in order to determine r, the symbol for the correlation coefficient.

One can plot the relationship between two measures by use of a scattergram. Figures 4 to 6 present three degrees of relationship, or correlation, which might be found for two variables.

Figure 4 shows no correlation between X and Y. If these were the verbal and quantitative test scores and $r \approx 0$, we would say that there is no relationship between the two aptitudes. Figure 5 shows a nearly perfect

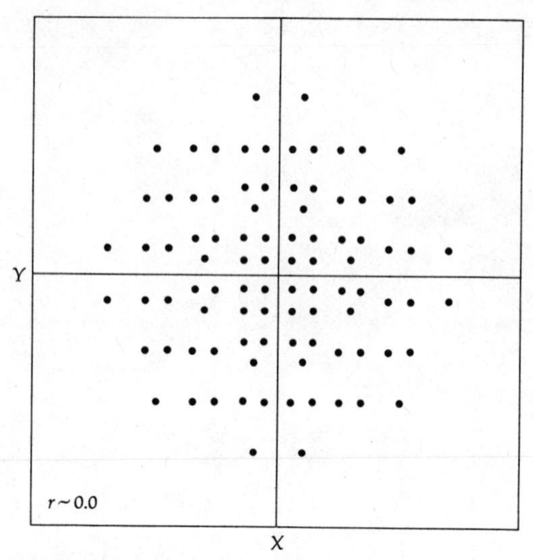

FIGURE 4

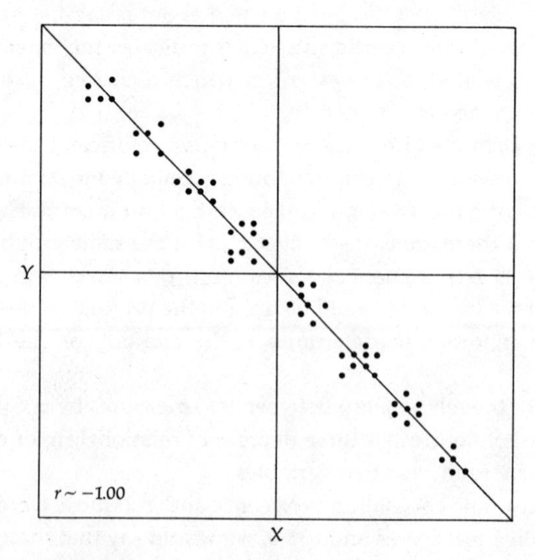

FIGURE 5

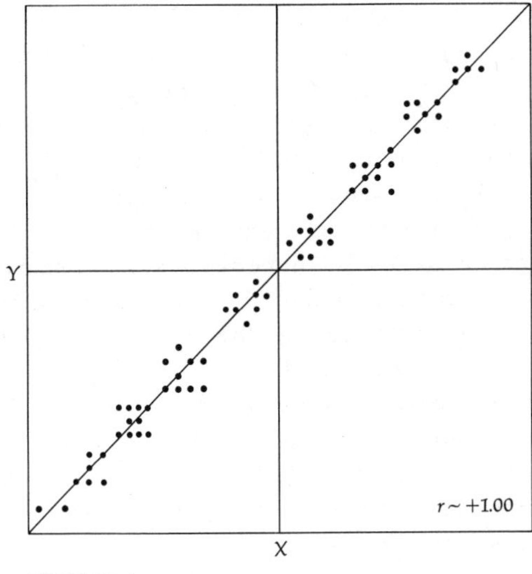

FIGURE 6

negative correlation, meaning that scores high on *X*, for example, verbal aptitude, tend to go with low scores on *Y*, for example, quantitative scores. The opposite is true for Fig. 6, where high scores on *X* match high *Y* scores, and low *X* scores match low *Y* scores.

A correlation can be between +1.00 to .00 to −1.00. Any correlation obtained between two variables does *not* imply that there is a cause-and-effect relationship, however. It merely indicates that as one variable varies, there is some degree of systematic change in the other variable, but one does not *cause* the other. For example, if the correlation between our sample of scores on the verbal and quantitative tests were .70, we would say that there seems to be a strong relationship between these two aptitudes, but we would *never* say that one *caused* the other. In other words, a high score on the verbal test did not cause a student to get a high score on the quantitative test. It is very important to remember this restriction on the interpretation of a correlation coefficient.

If one wished to go a little further, one could say that the correlation derived from the two variables can be used as an indication of the percent of variance accounted for by one of the variables. With a correlation of

.80 between X and Y, approximately .64, or 64 percent, of the variance in X scores can be accounted for by Y. This figure is derived by squaring the correlation coefficient, that is, $(.80)^2 = .64$, or 64 percent.

One can use the information obtained from the two variables, specifically, the means of the two and their standard deviations, in plotting linear equations. One can then use the line derived therefrom in predicting Y from X or X from Y. Again, a standard statistics book will explain the use of linear regression. The important thing to remember is that if you plot the points on a scattergram, the closer the points fall to a single line which might be drawn through the points, the higher will be the correlation between the two variables.

It would help your understanding of the concept of correlation to turn to exercise 2 of Appendix E and compute the coefficient for the sets of scores listed there. Begin by plotting the pairs of scores on a scattergram, and predict about what the correlation might be. Then use the scores in the formula to check your estimate.

APPENDIX D

Analysis-of-variance Approach to Reliability

Soon after the publication of the Kuder-Richardson formulas, Jackson (1939), Hoyt (1941), and Alexander (1947) published papers which proposed to use analysis-of-variance techniques to provide the same estimate of test reliability. The most widely used of these methods is that proposed by Hoyt. He has shown that the result obtained by his method is identical with that obtained with the Kuder-Richardson formula 20. The item data shown in Table 3 in the text may be considered to be a two-way factorial design for analysis-of-variance, without replications; i.e., each item is given to each student only once. Guilford (1954, p. 384) has interpreted the Hoyt formulas as follows:

$$\boxed{r_{tt} = 1 - \frac{V_r}{V_{\text{ex}}} = \frac{V_{\text{ex}} - V_r}{V_{\text{ex}}}}$$

Reliability as defined in analysis-of-variance terms

where V_r is the variance for the remainder sum of squares, and V_{ex} is the variance for the examinees. The sum of squares for the components of the formula may be computed as follows:

$$SS_{\text{ex}} = \frac{\Sigma X_t^2}{k} - \frac{(\Sigma X_t)^2}{kN}$$

where

SS_{ex} = sum of squares for examinees
X_t = total score for each examinee
k = number of test items
N = number of examinees

$$SS_{it} = \frac{\Sigma R_i^2}{N} - \frac{(\Sigma X_t)^2}{kN}$$

where

SS_{it} = sum of squares for items
R_i = number of correct responses for item i, and other symbols as defined above

$$SS_{tot} = \frac{\Sigma R_i \, \Sigma W_i}{\Sigma R_i + \Sigma W_i}$$

where

SS_{tot} = total sum of squares
W_i = number of wrong responses to item i, and other symbols as defined above

$$SS_r = SS_{tot} - SS_{ex} - SS_{it}$$

where

SS_r = sum of squares for the remainder

The item-by-student data in Table 3 will be used to illustrate the use of the analysis-of-variance estimate of reliability. The sum of squares for examinees is

$$SS_{ex} = \frac{135}{8} - \frac{(27)^2}{(8)(6)} = 16.88 - 15.19 = 1.69$$

The sum of squares for items is

$$SS_{it} = \frac{103}{6} - \frac{(27)^2}{(8)(6)} = 17.17 - 15.19 = 1.98$$

The total sum of squares is

$$SS_{tot} = \frac{(27)(21)}{(27) + (21)} = \frac{567}{48} = 11.81$$

The remainder sum of squares is

$$SS_{rem} = 11.81 - 1.69 - 1.98 = 8.14$$

These values can be placed in an analysis-of-variance table with their associated degrees of freedom. Degrees of freedom for examinees equals $N-1$; degrees of freedom for items equals $k-1$; degrees of freedom for remainder equals the degrees of freedom for examinees times the degrees of freedom for items.

Source of Variance	*Sum of Squares*	*Degrees of Freedom*	*Variance*
Examinees	1.69	5	.338(V_{ex})
Items	1.98	7	.283(V_{it})
Remainder	8.14	35	.233(V_{rem})
Total	11.81	47	

Using the basic Hoyt formula, $r_{tt} = 1 - V_{rem}/V_{ex}$, the analysis-of variance reliability estimate for the data in Table 3 is

$$r_{tt} = 1 - \frac{.233}{.338} = 1 - .672 = .328$$

The slight difference between this estimate and that obtained by the Kuder-Richardson formula 20 is due to rounding in the computations.

It should be noted that the variance of interest in the Hoyt approach in terms of estimating reliability is the variance among the examinees. From this quantity is subtracted that variance which is *not* accounted for by variation in the items, i.e., the remainder variance.

The Hoyt formula is similar to the theoretical conception of reliability.

$$\text{Hoyt reliability} = 1 - \frac{V_{rem}}{V_{ex}} \qquad \text{test theory reliability} = 1 - \frac{\sigma^2 \text{ error}}{\sigma^2 \text{ total}}$$

In the Hoyt formula the remainder variance is considered as error variance, and the variance of the examinees is considered the total variance which is of consequence.

The analysis-of-variance approach for estimating reliability is an internal-consistency measure. The same method may be used to analyze data from any single test which are used in test-retest and alternate forms reliability studies. The analysis-of-variance method may also be used to measure the reliability of stratified tests, i.e., tests in which there are clusters of items that are logically related to each other in some way. An example of a stratified test is a reading test in which students are required to read a paragraph, answer several questions, read another paragraph, answer more questions, etc. It would be predicted that there would be greater internal consistency among the questions asked about any one set of paragraphs as opposed to the internal consistency among all the items in the test. By using the stratified reliability procedures, one may estimate the effect of the stratification in the test data. For further descriptions of the use of these procedures, refer to the articles by Jackson (1939) and Alexander (1947) for test-retest and alternate forms reliability discussions and to a paper by Rabinowitz and Eikeland (1964) for a description of stratified reliability-estimation procedures.

For practice in the use of the analysis-of-variance technique for determining reliability, refer to exercise 5 in Appendix E.

APPENDIX E

Practice Exercises on Reliability

1. Computation of Variance and Standard Deviation

Shown below are two sets of scores. Using the formulas, compute the mean, variance, and standard deviation for each set. The answers follow.

$$\overline{X} = \frac{\Sigma X}{N} \qquad \sigma^2 = \frac{\Sigma X^2}{N} - \overline{X}^2 \qquad \sigma = \sqrt{\sigma^2}$$

SET 1	SET 2
9	43
7	26
8	35
11	44
14	33
10	38
9	35
8	46
12	29
	40

Answers to Exercise 1

SET 1	SET 2
$\overline{X} = \frac{88}{9} = 9.78$	$\overline{X} = \frac{369}{10} = 36.9$
$\sigma^2 = \frac{900}{9} - (9.78)^2$	$\sigma^2 = \frac{14{,}001}{10} - (36.9)^2$
$= 100 - 95.59$	$= 1{,}400.1 - 1{,}361.6$
$= 4.41$	$= 38.5$
$\sigma = 2.1$	$\sigma = 6.2$

2. Computation of a Product-Moment Correlation Coefficient

Listed below are two sets of pairs of test scores. Using the formula, compute the product-moment correlation coefficient for each set. The answers follow.

$$r = \frac{N\Sigma XY - \Sigma X \Sigma Y}{\sqrt{[N\Sigma X^2 - (\Sigma X)^2]\,[N\Sigma Y^2 - (\Sigma Y)^2]}}$$

SET 1		SET 2	
4	6	27	16
3	4	33	21
5	3	26	18
7	9	31	25
6	5	37	23
9	9	29	17
7	5	25	19
6	8	32	24
		28	16
		34	24

Answers to Exercise 2

SET 1

$N = 8 \qquad \Sigma XY = 308$

$\Sigma X = 47 \qquad \Sigma Y = 49$

$\Sigma X^2 = 301 \qquad \Sigma Y^2 = 337$

$$r = \frac{(8)(308) - (47)(49)}{\sqrt{[(8)(301) - (47)^2] - [(8)(337) - (49)^2]}}$$

$$= \frac{161}{\sqrt{58{,}705}} = .66$$

SET 2

$N = 10 \qquad \Sigma XY = 6{,}219$

$\Sigma X = 302 \qquad \Sigma Y = 203$

$\Sigma X^2 = 9{,}254 \qquad \Sigma Y^2 = 4{,}233$

$$r = \frac{(10)(6{,}219) - (302)(203)}{\sqrt{[(10)(9{,}254) - (302)^2]\,[(10)(4{,}233) - (203)^2]}}$$

$$= \frac{884}{\sqrt{1{,}497{,}656}} = .72$$

3. Computation of the Kuder-Richardson formula 20 Reliability Coefficient

Listed below is a matrix of 1's and 0's, which represent the performance of 10 students on a six-item test. Using the formula, compute the Kuder-Richardson formula 20 reliability coefficient for the test. The answer follows.

$$\text{K-R 20} = \left(\frac{K}{K-1}\right)\frac{\sigma_o^2 - \Sigma pq}{\sigma_o^2}$$

	Item						
Student	*1*	*2*	*3*	*4*	*5*	*6*	*Total*
1	1	1	1	0	1	1	5
2	0	1	1	0	0	1	3
3	1	0	1	1	0	0	3
4	1	1	0	1	1	1	5
5	1	1	1	1	0	0	4
6	0	1	0	0	1	0	2
7	1	1	0	1	1	0	4
8	1	1	0	1	0	1	4
9	1	1	1	1	1	1	6
10	1	1	1	1	1	1	6

Answer to Exercise 3

	Item					
	1	*2*	*3*	*4*	*5*	*6*
p	.8	.9	.6	.7	.6	.6
q	.2	.1	.4	.3	.4	.4
pq	.16	.09	.24	.21	.24	.24
Σpq	1.18					

$\Sigma X = 42 \qquad N = 10 \qquad \overline{X} = 4.2 \qquad K = 6$

$\Sigma X^2 = 192$

$\sigma^2 = 192/10 - (4.2)^2 = 19.20 - 17.64 = 1.56$

$$\text{K-R } 20 = \left(\frac{6}{5}\right)\left(\frac{1.56 - 1.18}{1.56}\right) = (1.2)\left(\frac{.38}{1.56}\right)$$

$$\text{K-R } 20 = .29$$

4. Computation of the Kuder-Richardson formula 21 Reliability Coefficient

Listed below are a set of total scores for a group of 20 students on a 50-item final examination. Compute the Kuder-Richardson formula 21 reliability by using the formula. As a second problem, compute the same coefficient for the total scores obtained by the 10 students in exercise 3. The answers follow.

$$\text{K-R 21} = \left(\frac{K}{K-1}\right)\frac{\sigma_o^2 - K\overline{pq}}{\sigma_o^2}$$

SET 1

43	44
37	43
41	46
46	44
47	42
42	40
43	39
39	41
47	47
43	38

Answers to Exercise 4

SET 1

$N = 20 \quad \Sigma X = 852 \quad \Sigma X^2 = 36{,}472$

$K = 50$

$\overline{X} = 852/20 = 42.6$

$\sigma^2 = 36{,}472/20 - (42.6)^2 = 8.84$

$\overline{p} = 42.6/50 = .852 \qquad \overline{q} = 1 - .852 = .148$

$$\text{K-R 21} = \left(\frac{50}{49}\right)\left[\frac{8.84 - (50)(.852)(.148)}{8.84}\right]$$

$$= (1.02)\left(\frac{8.84 - 6.30}{8.84}\right)$$

$$\text{K-R 21} = .293$$

SET 2 (Scores from Exercise 3)

$N = 10 \quad \overline{X} = 4.2 \quad \sigma^2 = 1.56 \quad K = 6$

$\overline{p} = 4.2/6 = .7 \qquad \overline{q} = 1.0 - .7 = .3$

$$\text{K-R 21} = \left(\frac{6}{5}\right)\left[\frac{1.56 - (6)(.7)(.3)}{1.56}\right] = (1.2)\left(\frac{.30}{1.56}\right)$$

$$\text{K-R 21} = .230$$

5. Computation of a Reliability Coefficient through the Analysis-of-variance

Using the formulas listed in Appendix D and the table of items for exercise 3, compute the reliability of the test by analysis-of-variance. The correct answer follows.

Answer to Exercise 5

	Item					
	1	2	3	4	5	6
R	8	9	6	7	6	6
R^2	64	81	36	49	36	36
W	2	1	4	3	4	4

$\Sigma X = 42$ $\quad N = 10$ $\quad \Sigma R^2 = 422$ $\quad \Sigma W = 18$

$\Sigma X^2 = 192$ $\quad K = 6$ $\quad \Sigma R = 302$

$$SS_{ex} = \frac{192}{6} - \frac{(42)^2}{60} = 32 - 29.4 = 2.6$$

$$SS_{it} = 302/10 - 29.4 = 30.2 - 29.4 = .8$$

$$SS_{tot} = \frac{(42)(18)}{42 + 18} = \frac{756}{60} = 12.6$$

$$SS_{rem} = 12.6 - 2.6 - .8 = 9.2$$

Source of Variance	Sum of Squares	Degree of Freedom	Variance
Examinees	2.6	9	.289
Items	.8	5	.160
Remainder	9.2	45	.204
Total	12.6	59	

$$r_{tt} = 1 - \frac{V_{rem}}{V_{ex}} = 1.0 - \frac{.204}{.289} = .294$$

6. Computation of the Reliability of Judges' Ratings

Using the rankings of three judges of eight individuals, compute the reliability of the data listed below. Use the Guilford and Spearman-Brown formulas. The answer follows.

$$\bar{r} = 1 - \frac{K(4N + 2)}{(K - 1)(N - 1)} + \frac{12\Sigma S^2}{K(K - 1)N(N^2 - 1)}$$

$$\text{Spearman-Brown} = \frac{nr_{11}}{1 + (n - 1)r_{11}}$$

Individual	*Judge*		
	1	*2*	*3*
1	4	3	2
2	5	2	1
3	1	4	5
4	8	6	7
5	6	7	6
6	2	1	3
7	3	5	4
8	7	8	8

Answer to Exercise 6

$K = 3 \qquad N = 8 \qquad \Sigma S^2 = 1{,}756$

$$\bar{r} = 1 - \frac{(3)(34)}{(2)(7)} + \frac{(12)(1{,}756)}{(3)(2)(8)(63)}$$

$= 1 - 7.285 + 6.968$

$= .683$ Reliability of an individual judge

Spearman-Brown reliability for three judges:

$$SB_{rel} = \frac{(3)(.683)}{1 + (2)(.683)} = \frac{2.049}{2.366} = .866$$

APPENDIX F

Test Questions and Answers for Reliability

Reliability Final Examination

1. The ______________ of a set of test scores indicates either the stability or consistency of those scores.

2. Standard deviation is defined as the ______________ of the variance.

3. The theoretically exact measure of a student's ability to perform on a test is referred to as his ______________.

4. The major assumption about the errors associated with test scores is that they are ______________ distributed.

5. Reliability is defined as the ratio of the ______________ variance in a set of scores to the ______________ variance.

6. When the same test is administered twice in order to determine the reliability of the scores, the ______________ method for determining reliability has been used.

7. The factor of "memory" contributes to ______________ variance with the test-retest method but contributes to ______________ variance with the alternate forms method.

8. Internal-consistency reliability estimates are based upon ________ *(how many)* administration(s) of a test.

9. When the split-half method is used to estimate the reliability of a test, it is necessary first to compute the correlation between the part scores, and then to employ the ______________ formula.

10. The Kuder-Richardson formula 20 reliability estimate is the equivalent to all possible ______________ estimates.

11. The reliability coefficient obtained by the use of Kuder-Richardson formula 21 will be a ______________ estimate of reliability when compared with the more exact K-R 20 estimate; it will be a ______________ *(conservative/* ______________ estimate of a test's reliability. *liberal)*

12. Since items which are measuring the same skill are more likely to have ______________ interitem correlations than items which measure different skills, higher reliabilities should be expected from ______________ factor tests.

13. The smaller the standard error of measurement, the ______________ the confidence one can have in any single obtained score.

14. If the reliability of a test is to be increased by adding more items, these items should have the same ______________ and ______________.

15. If 15 percent of a group of students complete a test (which is appropriate for them), it would be considered a ______________ test; if 85 percent complete it, it would be considered a ______________ test.

16. If the range of abilities in a test group is wider than that intended by the test maker, the reliability estimate will be spuriously ______________, while if the range of abilities is significantly restricted, the reliability estimate will be spuriously ______________.

17. In the text it was pointed out that the standard error of measurement could be interpreted for the individual much as the ______________ can be interpreted for a group.

18. Test items with a difficulty of ______________ will maximize the variance among the students; however, such items may not provide the ______________ among students that is desired.

19. In order to have maximally reliable differences in subtest scores, the subtests should be highly ______________ and ______________.

20. The result of applying Guilford's formula to a set of ratings is the reliability of ______________ rater(s).

21. In order to assess the reliability of the constancy of a trait over a period of time, it would be most suitable to employ the ____________ method for estimating reliability.

22. Professionally developed standardized tests often have reported reliabilities above ____________, while classroom tests often have reliabilities of ____________ or below.

23. The test user must be aware that a test may have ____________ reliability coefficient(s).

24. In addition to test reliability, the standard ____________ should be reported.

Identify the following formulas:

25. $\left(\frac{K}{K-1}\right)\frac{\sigma_t^2 - \Sigma pq}{\sigma_t^2}$ ____________

26. $1 - \frac{\sigma_e^2}{\sigma_t^2}$ ____________

27. $\frac{\Sigma X^2}{N} - \overline{X}^2$ ____________

28. $\sigma_t\sqrt{1 - r_{tt}}$ ____________

QUESTIONS 29–34: Four formulas are presented. Below each formula is one or more symbols. Select from the following list the correct identification of each of the symbols.

a. Number of items
b. Number of students
c. reliability
d. validity
e. mean proportion correct
f. mean proportion incorrect
g. ratio of new test relative to old
h. correlation between items a and b
i. number correct
j. standard deviation
k. reliability of items a and b

$$\text{Spearman-Brown formula} = \frac{nr_{11}}{1 + (n-1)r_{11}}$$

29. $n =$ ______________

$$\text{Kuder-Richardson formula 21} = \left(\frac{K}{K-1}\right)\frac{\sigma_t^2 - k\overline{pq}}{\sigma_t^2}$$

30. $K =$ ______________

31. $\overline{p} =$ ______________

$$\text{Standard error of measurement} = \sigma_t\sqrt{1 - r_{tt}}$$

32. $r_{tt} =$ ______________

33. $\sigma_t =$ ______________

$$\text{Reliability of difference between two standard scores} = \frac{r_{aa} + r_{bb} - 2r_{ab}}{2 - 2r_{ab}}$$

34. $r_{ab} =$ ______________

QUESTIONS 35–38: These problems require the use of the formulas which are listed in questions 29 to 34. Refer to those formulas to solve the following problems.

35. Compute the Kuder-Richardson formula 21 reliability of a set of scores from a 30-item test which has a mean of 24 and a variance of 8.

$$\left(\overline{p} = \frac{\overline{X}}{K};\ \overline{q} = 1 - \overline{p}\right)$$

36. What is the standard error of measurement for a set of scores which have a reliability of .75 and a standard deviation of 7? (Be careful to enter the decimal point accurately.)

37. If a 30-item test, with a reliability of .64, were increased in length to 60 items, what would be the estimated reliability of the new test according to the Spearman-Brown formula?

38. Two subtests on a particular achievement test have reported reliabilities of .55 and .70, and the correlation between scores on the two subtests is .40. What is the reliability of the difference between scores on the two subtests?

ESSAY QUESTIONS

39. Explain the meaning of the expression $X_o = X_t + X_e$ and discuss the three assumptions of test theory.

40. List the three major types of reliability coefficients and indicate uses and precautions associated with each.

41. Describe the usefulness in reporting the standard error of measurement as well as the reliability for a test.

42. Discuss the question, "How reliable should a test be?"

43. List at least five questions that should be taken into account when considering the reported reliability of a standardized test.

44. Consider the following hypothetical situation: You have just acquired funds to test your fifth-grade students in reading and arithmetic. The reliability of one of the tests you are considering buying reads as follows:

> Reliability coefficients for the XYZ Achievement Test are based upon a sample of children from New York and Nebraska. All students were in grades 3, 4 and 5. The test-retest reliability of the test is .85.

As a potential consumer, criticize this report, and raise other questions about the reliability of the test which you would like to have answered before you purchased the test.

Answers to Reliability Test

1. Reliability
2. Square root
3. True score
4. Randomly
5. True-score, total-score
6. Test-retest
7. True-score, error
8. One

9. Spearman-Brown

10. Split-half

11. Lower, conservative

12. Higher, uni-

13. Greater (or higher, more, larger)

14. Content, psychometric characteristics (i.e., difficulty and interitem correlation)

15. Speeded, power

16. High, low

17. Standard deviation

18. .50, discrimination (or variability)

19. Reliable, uncorrelated

20. One

21. Test-retest

22. .90 (or between .80 and 1.00), .50 (or between .40 and .60)

23. Several (or more than one, many)

24. Error of measurement

25. Kuder-Richardson formula 20

26. Test-theory definition of reliability

27. Total or obtained variance

28. Standard error of measurement

29. g

30. a

31. e

32. c

33. j

34. h

35. $$\text{K-R 21} = \left(\frac{30}{30-1}\right)\left[\frac{8-(30)(\frac{24}{30})(1-\frac{24}{30})}{8}\right]$$

$$= \left(\frac{30}{29}\right)\left[\frac{8-(30)(.8)(.2)}{8}\right] = (1.036)\left[\frac{8-(30)(.16)}{8}\right]$$

$$= (1.036)\left(\frac{8-4.8}{8}\right) = (1.036)\left(\frac{3.2}{8}\right) = (1.036)(.4)$$

$$= .414$$

36. $$\sigma_t\sqrt{1-r_{tt}} = 7\sqrt{1-.75}$$

$$= 7\sqrt{.25}$$

$$= (7)(.5)$$

$$= 3.5$$

37. $$\frac{nr_{11}}{1+(n-1)r_{11}} = \frac{(\frac{60}{30})(.64)}{1+(\frac{60}{30}-1)(.64)} = \frac{(2)(.64)}{1+(2-1)(.64)}$$

$$= \frac{1.28}{1+(.64)} = \frac{1.28}{1+.64} = \frac{1.28}{1.64} = .78$$

38. $$r_{\text{diff}} = \frac{.55+.70-(2)(.40)}{2-(2)(.40)} = \frac{.45}{1.2} = .375$$

39. X_o = obtained score
X_t = true score
X_e = error score

Assumptions of test theory:

1. The true score is a stable quantity.
2. The error score is a random variable.
3. Any test score is composed of a true score and an error score.

40.

Type	*Uses*	*Precautions (Source of Error)*
Test-retest	Power or speeded tests; determine stability of content	Temporary conditions affecting S's memory of specific knowledge; conditions of test administration
Alternate form	Power or speeded test; determine stability of student performance	Temporary conditions affecting S; conditions of test administration
Split-half	Power test; determine consistency within a test	Temporary conditions affecting S
		OR: Understand the logic of the study, nature of the coefficients, and the conditions which affect them

41. SE_m indicates the amount of confidence one can place in an individual's test score. Using the r_{tt} and SE_m of the normative group, one determines the confidence he can place in using the test on another sample: whether it is appropriate for the designated group.

42. Reliability should be as high as possible, but depends on the importance of the situation for which the test is to be used. Critical decisions demand high reliability; routine group decisions can accept $r_{tt} = .50$; for decisions where no other identification is possible, and incorrect classification is not detrimental to the individuals, a lower r_{tt} may be acceptable.

43. Possible questions:

1. For whom is the test intended?
2. On whom was the test standardized?
3. How well do items 1 and 2 compare?
4. How was reliability computed (method)?
5. What use is to be made of the scores?
6. Is the r_{tt} sufficient for the decision to be made based on test scores?

44. Possible comments:

1. What does XYZ measure? Achievement in what?

2. On whom was the test originally standardized, if anyone?
3. How many students were in the Nebraska sample? By grade?
4. What is the SE_m for this group?
5. Are there separate r_{tt} and SE_m reports for each grade (third, fourth, and fifth)?
6. What was the test-retest time interval? Is it speeded or power?

APPENDIX G

Test Questions and Answers for Validity

Validity Final Examination

1. If one asks the general question, "Does the test measure what I want it to measure?", he is referring to the ______________ of the test.

2. To determine a validity coefficient for a test, scores are correlated with a ______________.

3. The type of validity which relates a person's present performance with future performance is ______________ validity.

4. When one determines the validity of one instrument by correlating it with another performance which occurs at approximately the same point in time, he is establishing ______________ validity.

5. A high validity coefficient implies that one can use the instrument as a good predictor of either ______________ or ______________ performances.

6. Validity is determined by giving the test to a ______________ sample of subjects.

7. The symbol r_{xx} indicates the correlation between a ______________ and ______________.

8. The symbol r_{xy} indicates the correlation between a ______________ and ______________.

9. The extent to which the items in a test sample a subject area or course is termed ________________ validity.

10. Construct validity is most often related to the testing of a __________.

11. By examining the items in a test, but not testing them, one is determining the ________________ validity of the test.

12. Restricting the ability range of a group ________________ the variance.

13. The ________________ of a test cannot exceed the square root of the ________________ of a test.

14. A test whose items are samples of a particular situation or behavior is said to have ________________ validity.

15. Test X has been developed to measure the reading comprehension of grade school children. When given to a representative group, the reliability was .72. To determine the validity of the test, the best group to use would be:

a. 10 students in each grade from K through high school
b. 100 students each in first, fifth, and eighth grade
c. 300 average students
d. All students in the grade school where the test is to be used

16. The validity estimate of the test mentioned above will be highest if it is validated on:

a. All fifth-graders
b. All fourth-, fifth-, and sixth-graders
c. All students in grades 1 through 8
d. Only those students who can read

QUESTIONS 17–22: Below are situations you might encounter when analyzing data. Give the most appropriate correlation coefficient to use in order to estimate the validity of the test in question.

	Test	Criterion	Correlation
17.	Test questions: 1 = correct 0 = wrong	Male-female	________
18.	Sequential order in which 10 students enter the testing room	Grades on mid-term exam in statistics	________
19.	Grades on algebra test from 59 to 98	Male-female	________
20.	Pass-fail mark on algebra test, with 65 or below failing	Admittance to college, with GRE score of 500 or more passing	________
21.	Total score on home economics final exam	Total score on information test about food processing	________
22.	Scores on multiple-choice test, range from 65 to 127, in science	Passing or failing grade in science: A, B, C, D = pass, F = fail	________

QUESTIONS 23–30: Identify each of the following formulas:

23. $$\frac{N\Sigma XY - \Sigma X \Sigma Y}{\sqrt{[N\Sigma X^2 - (\Sigma X)^2]\,[N\Sigma Y^2 - (\Sigma Y)^2]}}$$ ________

24. $$\frac{(a)(d)}{(b)(c)}$$ ________

25. $$\frac{p_{ij} - p_i p_j}{\sqrt{(p_i)(q_i)(p_j)(q_j)}}$$ ________

26. $$\frac{\overline{X}_p - \overline{X}_t}{\sigma_t}\sqrt{\frac{p}{q}}$$ ________

27. $$\frac{\overline{X}_p - \overline{X}_t}{\sigma_t}\left(\frac{p}{z}\right)$$ ________

28. $$1 - \frac{6\Sigma D^2}{N(N^2 - 1)}$$ ________

29. $\frac{r_{xy}}{\sqrt{r_{yy}}}$ Correction for attenuation in ______________

30. $\frac{r_{xy}}{\sqrt{r_{xx} r_{yy}}}$ Correction for attenuation in ______________

PROBLEMS

31. Given two dichotomous variables:

X	0	0	1	1	0	1	0	0	1	0
Y	1	0	1	0	0	0	1	0	1	1

Find the validity for X based on criterion Y, given the formula

$$\text{phi} = \frac{p_{ij} - p_i p_j}{\sqrt{(p_i)(q_i)(p_j)(q_j)}}$$

a. $p_i =$ ______________ d. $q_i =$ ______________
b. $p_j =$ ______________ e. $q_j =$ ______________
c. $p_{ij} =$ ______________ f. phi = ______________

32. Given one continuous variable, the other a dichotomy:

X	12	13	10	9	11
Y	0	0	1	1	1

Complete the following:

a. $N =$ ______________

b. $\overline{X}_p =$ ______________

c. $\overline{X}_t =$ ______________

$\sigma_t = 1.4$

d. $p =$ ______________

e. What formula do you use to determine the validity of Test X using Y as criterion? ______________ Why? ______________

Biserial *Point-biserial*

$$r_{\text{bis}} = \frac{\overline{X}_p - \overline{X}_t}{\sigma_t}\left(\frac{p}{z}\right) \qquad r_{\text{pbis}} = \frac{\overline{X}_p - \overline{X}_t}{\sigma_t}\sqrt{\frac{p}{q}}$$

where Z = .399.

f. Find the validity coefficient for the formula designated in **e.**

r_{bis} = ______________ r_{pbis} = ______________

33. $r_{xx} = .64$. r_{xy} can be as high as ______________.

34. Given:

$$r_{xx} = .49$$
$$r_{yy} = .64$$
$$r_{xy} = .68$$

$$r_{cc} = \frac{r_{xy}}{\sqrt{r_{yy}}}$$

Correct for attenuation in the criterion only. What is the estimate of the validity of the test, if the criterion were perfect?

r_{cc} = ______________

35. Given:

		Q-1		
		Yes	No	Total
Q-2	Yes	30	20	40
	No	40	10	50
	Total	70	30	100

Both variables are continuous data from the following scale:

Absolutely	*Sometimes*	*Maybe*	*Rarely*	*Never*
	Yes		No	

a. What coefficient would you use to obtain a validity estimate for the above information? ______________

b. What coefficient do you get using

$$\frac{ad}{bc} = \_\_\_\_$$

c. What is the next step in estimating this correlation? ______________

36. Given:

Student	Test	Criterion
1	37	P
2	42	P
3	13	F
4	20	P
5	31	F

The criterion is the final grade received.

a. Indicate the coefficient you would use to determine the validity of the test. ______________

b. What is your rationale for using that one? ______________.

c. Without working the problem, examination of the data shows the correlation to be about:

(1) .95
(2) .50
(3) .30
(4) .10

ESSAY QUESTIONS

37. Discuss "cross-validation" in terms of its rationale and procedures.

38. Discuss "correction for attenuation" in terms of what it is, what it does, and why one would use it.

39. Set up a hypothetical situation in which you have Test X and want to be able to report as accurate a validity coefficient as you can. Define the variables, situation, formulas, and so forth.

40. Set up dummy data for question 39 and give the validity coefficient. Is that the highest estimate you can make?

What do the data tell you?

Answers to Validity Test

1. Validity
2. Criterion
3. Predictive
4. Concurrent
5. Present, future
6. Representative
7. Test, itself
8. Test, criterion
9. Curricular
10. Theory
11. Face
12. Restricts (decreases, limits, lowers)
13. Validity, reliability
14. Content
15. d
16. c
17. Phi, or ϕ
18. Rank order, or rho
19. Point-biserial
20. Tetrachoric
21. General product-moment
22. Biserial
23. General
24. Tetrachoric
25. Phi
26. pbis
27. bis
28. Rho, or rank order
29. Criterion only
30. Criterion and test
31. **a.** .4 **b.** .5 **c.** .2 **d.** .6 **e.** .5

f. $\text{phi} = \dfrac{.2 - (.4)(.5)}{\sqrt{(.4)(.6)(.5)(.5)}} = \dfrac{.2 - .2}{\sqrt{(.24)(.25)}} = \dfrac{0.0}{\sqrt{.06}} = 0.0$

32. $r_{\text{bis}} = \left(\dfrac{10 - 11}{1.4}\right)\left(\dfrac{.6}{.399}\right)$ $\qquad r_{\text{pbis}} = \dfrac{10 - 11}{1.4}\sqrt{\dfrac{.6}{.4}}$

$= \left(\dfrac{-1}{1.4}\right)(1.5)$ $\qquad = \dfrac{-1}{1.4}\sqrt{1.5}$

$= (-.71)(1.5)$ $\qquad = (-.71)(1.23)$

$= -1.065$ $\qquad = -.87$

$N = 5,\ \overline{X}_p = 10,\ \overline{X}_t = 11,\ p = .6$

33. $r_{xy} = .80$

34. $r_{cc} = \dfrac{.68}{\sqrt{.64}} = \dfrac{.68}{.8} = .85$

35. a. Tetrachoric $\quad \dfrac{ad}{bc} = \dfrac{(30)(10)}{(20)(40)} = \dfrac{300}{800} = \text{reverse}$

b. $\dfrac{(20)(40)}{(30)(10)} = \dfrac{800}{300} = 2.67$

c. Look up r_{tet} estimate. 2.67 is used to enter a table of ad/bc values for r_{tet}.
2.67 = .36

36. Depends:
pbis = continuous and dichotomous variables }
bis = continuous and forced dichotomy } .50

37. 1. Last validation of test.
2. Different sample, similar characteristics as first sample.
3. Usually results in a lower, more accurate estimate of validity.

38. 1. Correct for error variance in scores and attenuation in correlation coefficient.
2. Indicates what a correlation may be if one or both variables were perfect measures.
3. To decide whether or not one should take time to increase the reliability of the test.

39. Answer should include:

1. Whether variables are continuous or dichotomous
2. What group, when and where tested
3. Criterion used
4. Validation study, group, etc.
5. What formula

40. Answer should include:

1. Data
2. Use of formula
3. Correction for attenuation
4. $\sqrt{r_{xx}}$
5. How well x is predictive of y

REFERENCES

Alexander, H. W.: The Estimation of Reliability When Several Trails Are Available, *Psychometrika*, vol. 12, pp. 79–99, 1947.

Anastasi, Anne: "Psychological Testing," The Macmillan Company, New York, 1961.

Cronbach, L. J.: "Essentials of Psychological Testing," Harper & Row, Publishers, Incorporated, New York, 1960.

———and Goldine C. Gleser: The Signal/Noise Ratio in the Comparison of Reliability Coefficients, *Educ. Psychol. Meas.*, vol. 24, pp. 467–480, 1964.

Ebel, R. L.: Estimation of the Reliability of Ratings, *Psychometrika*, vol. 16, pp. 407–424, 1951.

Freeman, F. S.: "Theory and Practice of Psychological Testing," Holt, Rinehart and Winston, Inc., New York, 1963.

Guilford, J. P.: "Psychometric Methods," 2d ed., McGraw-Hill Book Company, New York, 1954.

———"Fundamental Statistics in Psychology and Education," 4th ed., McGraw-Hill Book Company, New York, 1965.

Hollingworth, H. L.: "Experimental Studies in Judgment," Science Press, New York, 1913.

Horst, P.: "Psychological Measurement and Prediction," Wadsworth Publishing Co., Inc., Belmont, Calif., 1966.

Hoyt, C.: Test Reliability Estimated by Analysis of Variance, *Psychometrika*, vol. 6, pp. 153–160, 1941.

Jackson, R. W. B.: Reliability of Mental Tests, *Brit. J. Psychol., Gen. Sect.*, vol. 29, pp. 267–287, 1939.

Kelley, T. L.: "Interpretation of Educational Measurements," World Book Company, Tarrytown-on-Hudson, N.Y., 1927.

Kuder, G. F., and M. W. Richardson: The Theory of Estimation of Test Reliability, *Psychometrika,* vol. 2, pp. 151–160, 1937.

McNemar, Quinn: "Psychological Statistics," John Wiley & Sons, Inc., New York, 1962.

Rabinowitz, W., and H. M. Eikeland: Estimating the Reliability of Tests with Clustered Items, *Pedagogisk Forskning,* 1964, pp. 85–106.

Rulon, P. J.: A Simplified Procedure for Determining the Reliability of a Test by Split-halves, *Harvard Educ. Rev.,* vol. 9, pp. 99–103, 1939.

Tyler, Leona E.: "The Psychology of Human Differences," Appleton-Century-Crofts, Inc., New York, 1965.

Wesman, A. G.: Reliability and Confidence, *Test Service Bull.* 44, pp. 2–7, The Psychological Corp., 1952.

BIBLIOGRAPHY

Barnette, W. Leslie, Jr. (ed.): "Readings in Psychological Tests and Measurements," The Dorsey Press, Homewood, Ill., 1964.

Buros, Oscar K. (ed.): "The Sixth Mental Measurements Yearbook," Gryphon Press, Highland Park, N.J., 1965.

Chase, Clinton L., and H. Glenn Ludlow (eds.): "Readings in Educational and Psychological Measurement," Houghton Mifflin Company, Boston, 1966.

Cronbach, Lee J.: "Essentials of Psychological Testing," Harper & Row, Publishers, Incorporated, New York, 1960.

Ghiselli, Edwin: "Theory of Psychological Measurement," McGraw-Hill Book Company, New York, 1964.

Gronlund, Norman E. (ed.): "Readings in Measurement and Evaluation," The Macmillan Company, New York, 1968.

Guilford, J. P.: "Fundamental Statistics in Psychology and Education," 4th ed., McGraw-Hill Book Company, New York, 1965.

Gulliksen, Harold: "Theory of Mental Tests," John Wiley & Sons, Inc., New York, 1950.

Helmstadter, G. C.: "Principles of Psychological Measurement," Meredith Publishing Company, Des Moines, 1964.

Jackson, Douglas N., and Samuel Messick: "Problems in Human Assessment," McGraw-Hill Book Company, New York, 1967.

Lindeman, Richard H.: "Educational Measurement," Scott, Foresman and Company, Glenview, Ill., 1967.

Lindquist, E. F. (ed.): "Educational Measurement," American Council on Education, Washington, D.C., 1951.

Payne, David A., and Robert F. McMorris: "Educational and Psychological Measurement," Blaisdell Publishing Company, Waltham, Mass., 1967.

Standards for Educational and Psychological Tests and Manuals, American Psychological Association, Washington, D.C., 1966.

Thorndike, Robert L., and Elizabeth Hagen: "Measurement and Evaluation in Psychology and Education," John Wiley & Sons, Inc., New York, 1955.

INDEX

INDEX